24 0044420

MW00592972

RUSSIAN TRANSLATIONS SERIES

1. K.Ya. Kondrat'ev et al. (editors): USSR/USA *Bering Sea Experiment*
2. D.V. Nalivkin: *Hurricanes, Storms and Tornadoes*
3. V.M. Novikov (editor): *Handbook of Fishery Technology*, Volume 1
4. F.G. Martyshev: *Pond Fisheries*
5. R.N. Burukovskii: *Key to Shrimps and Lobsters*
6. V.M. Novikov (editor): *Handbook of Fishery Technology*, Volume 4
7. V.P. Bykov (editor): *Marine Fishes*
8. N.N. Tsvelev: *Grasses of the Soviet Union*
9. L.V. Metlitskii et al.: *Controlled Atmosphere Storage of Fruits*
10. M.A. Glazovskaya: *Soils of the World* (2 volumes)
11. V.G. Kort & V.S. Samoilenko: *Atlantic Hydrophysical Polygon-70*
12. M.A. Margdzhanishvili: *Seismic Design of Frame-panel Buildings and Their Structural Members*
13. E.'A. Sokolenko (editor): *Water and Salt Regimes of Soils: Modeling and Management*
14. A.P. Bocharov: *A Description of Devices Used in the Study of Wind Erosion of Soils*
15. E.S. Artsybashev: *Forest Fires and Their Control*
16. R.Kh. Makasheva: *The Pea*
17. N.G. Kondrashova: *Shipboard Refrigeration and Fish Processing Equipment*
18. S.M. Uspenskii: *Life in High Latitudes*
19. A.V. Rozova: *Biostratigraphic Zoning and Trilobites of the Upper Cambrian and Lower Ordovician of the Northwestern Siberian Platform*
20. N.I. Barkov: *Ice Shelves of Antarctica*
21. V.P. Averkiev: *Shipboard Fish Scouting and Electronavigational Equipment*
22. D.F. Petrov (Editor-in-Chief): *Apomixis and Its Role in Evolution and Breeding*
23. G.A. Mchedlidze: *General Features of the Paleobiological Evolution of Cetacea*
24. M.G. Ravich et al.: *Geological Structure of Mac. Robertson Land (East Antarctica)*
25. L.A. Timokhov (editor): *Dynamics of Ice Cover*
26. K.Ya. Kondrat'ev: *Changes in Global Climate*
27. P.S. Nartov: *Disk Soil-Working Implements*
28. V.L. Kontrimavichus (Editor-in-Chief): *Beringia in the Cenozoic Era*
29. S.V. Nerpin & A.F. Chudnovskii: *Heat and Mass Transfer in tthe Plant-Soil-Air System*
30. T.V. Alekseeva et al.: *Highway Machines*
31. N.I. Klenin et al.: *Agricultural Machines*
32. V.K. Rudnev: *Digging of Soils by Earthmovers with Powered Parts*
33. A.N. Zelenin et al.: *Machines for Moving the Earth*
34. *Systematics, Breeding and Seed Production of Potatoes*
35. D.S. Orlov: *Humus Acids of Soils*
36. M.M. Severnev (editor): *Wear of Agricultural Machine Parts*
37. Kh.A. Khachatryan: *Operation of Soil-Working Implements in Hilly Regions*
38. L.V. Gyachev: *Theory of Surfaces of Plow Bottoms*
39. S.V. Kardashevskii et al.: *Testing of Agricultural Technological Processes*
40. M.A. Sadovskii (editor): *Physics of the Earthquake Focus*
41. I.M. Dolgin: *Climate of Antarctica*
42. V.V. Egorov et al.: *Classification and Diagnostics of Soils of the USSR*

RUSSIAN TRANSLATIONS SERIES (Contd.)

43. V.A. Moshkin: *Castor*
44. E.'T. Sarukhanyan: *Structure and Variability of the Antartic Circumpolar Current*
45. V.A. Shapa (Chief editor): *Biological Plant Protection*
46. A.I. Zakharova: *Estimation of Seismicity Parameters Using a Computer*
47. M.A. Mardzhanishvili & L.M. Mardzhanishvili: *Theoretical and Experimental Analysis of Members of Earthquake-proof Frame-panel Buildings*
48. S.G. Shul'man: *Seismic Pressure of Water on Hydraulic Structures*
49. Yu.A. Ibad-zade: *Movement of Sediments in Open Channels*
50. I.S. Popushoi (Chief editor): *Biological and Chemical Methods of Plant Protection*
51. K.V. Novozhilov (Chief editor): *Microbiological Methods for Biological Control of Pests of Agricultural Crops*
52. K.I. Rossinskii (editor): *Dynamics and Thermal Regimes of Rivers*
53. K.V. Gnedin: *Operating Conditions and Hydraulics of Horizontal Settling Tanks*
54. G.A. Zakladnoi & V.F. Ratanova: *Stored-grain Pests and Their Control*
55. Ts.E. Mirtskhulava: *Reliability of Hydro-reclamation Installations*
56. Ia. S. Ageikin: *Off-the-road Mobility of Automobiles*
57. A.A. Kmito & Yu.A. Sklyarov: *Pyrheliometry*
58. N.S. Motsonelidze: *Stability and Seismic Resistance of Buttress Dams*
59. Ia.S. Ageikin: *Off-the-road Wheeled and Combined Traction Devices*
60. Iu.N. Fadeev & K.V. Novozhilov: *Integrated Plant Protection*
61. N.A. Izyumova: *Parasitic Fauna of Reservoir Fishes of the USSR and Its Evolution*
62. O.A. Skarlato (Editor-in-Chief): *Investigation of Monogeneans in the USSR*
63. A.I. Ivanov: *Alfalfa*
64. Z.S. Bronshtein: *Fresh-water Ostracoda*
65. M.G. Chukhrii: *An Atlas of the Ultrastructure of Viruses of Lepidopteran Pests of Plants*
66. E.S. Bosoi et al.: *Theory, Construction and Calculations of Agricultural Machines*, Volume 1
67. G.A. Avsyuk (Editor-in-Chief): *Data of Glaciological Studies*
68. G.A. Mchedlidze: *Fossil Cetacea of the Caucasus*
69. A.M. Akramkhodzhaev: *Geology and Exploration of Oil- and Gas-bearing Ancient Deltas*
70. N.M. Berezina & D.A. Kaushanskii: *Presowing Irradiation of Plant Seeds*
71. G.U. Lindberg & Z.V. Krasyukova: *Fishes of the Sea of Japan and the Adjacent Deltas*
72. N.I. Plotnikov & I.I. Roginets: *Hydrogeology of Ore Deposits*
73. A.V. Balushkin: *Morphological Bases of the Systematics and Phylogeny of the Nototheniid Fishes*
74. E.Z. Pozin et al.: *Coal Cutting by Winning Machines*
75. S.S. Shul'man: *Myxosporidia of the USSR*
76. G.N. Gogonenkov: *Seismic Prospecting for Sedimentary Formations*
77. I.M. Batugina & I.M. Petukhov: *Geodynamic Zoning of Mineral Deposits for Planning and Exploitation of Mines*
78. I.I. Abramovich & I.G. Klushin: *Geodynamics and Metallogeny of Folded Belts*
79. M.V. Mina: *Microevolution of Fishes*
80. K.V. Konyaev: *Spectral Analysis of Physical Oceanographic Data*
81. A.I. Tseitlin & A.A. Kusainov: *Role of Internal Friction in Dynamic Analysis of Structures*
82. E.A. Kozlov: *Migration in Seismic Prospecting*
83. E.S. Bosoi et al.: *Theory, Construction and Calculations of Agricultural Machines, Volume 2*
84. B.B. Kudryashov and A.M. Yakovlev: *Drilling in the Permafrost*
85. T.T. Klubova: *Clayey Reservoirs of Oil and Gas*

CLAYEY RESERVOIRS OF OIL AND GAS

Clayey Reservoirs of Oil and Gas

T.T. KLUBOVA

RUSSIAN TRANSLATIONS SERIES
85

1991
A.A. BALKEMA/ROTTERDAM

Translation of: *Glinistie Kollectori nefti i gaza*, Nedra, Moscow, 1988.

Translator : S. Viswanathan
Technical Editor : P.K. Bose
Language Editor : Margaret Majithia

ISBN 90 6191 992 4

Preface

From 1912 through 1918 an intensive effort was directed towards accelerating the tempo of oil and condensate gas production. This led to opening new fields and new horizons in known ancient oil occurrences. One of the essential additional studies in oil-finding involved determination of the hydrocarbons present in the clayey reservoirs of the Volga-Ural region, the Carpathians, eastern Cis-Caucasus and other well-established oil-fields.

Clayey sedimentary rocks are widely distributed in the stratosphere. According to the estimates of A.E. Fersman, they comprise 80% of the sedimentary cover of the earth, which amounts to nearly 4% of the mass of the earth's crust. Hence, it was but natural that specialists from all fields of science long ago studied in detail not only those clayey rocks associated with economic minerals, but even considered the rocks *per se* as important economic mineral deposits. Thus studies were not confined to oil-bearing regions alone. The development of the oil industry world-wide had only just begun, with studies on the conditions of formation of oil and associated deposits and clayey rocks and their genesis gradually assuming greater significance.

Subsequently, multitudinous investigations established that the entire natural history of oil is related to the quantity of volcanic material and the depth of the basin has been established.

The capacity and filtration parameters of the clayey reservoirs of the Menilitovian formation, determined on core samples for example, correlate with those obtained for other clayey reservoirs. Porosity is around 5% and permeability less than $10^{-3} \mu m^2$. It is but natural that for these rocks the dimensions and extension of the weakened zones, and not just porosity and permeability, decide their reservoir potential.

Primarily the clayey composition of the Lower Menilitovian member, with OM and silica in rock-forming proportions, has given rise to the textural heterogeneity in which the textures at the microlevel contribute to the basic capacity of these rocks. Mesotextures play only a secondary role. The mechanism of formation of textural heterogeneity, and that means of the weakened zones, is similar to the one described for the rocks of the Bazhenovian formation. The parameters of the pore space were found to be very close. The average size of the chord of the pores varies from 0.56–2.4 μm. The ratio of length of the weakened zones to unit area is 1.95–2.21 mm/mm^2. The coefficient of orientation is 1.325.

According to geophysical and deep drilling data, transverse uplifts and depressions developed in the basement of the Outer and Inner zones of the Cis-Carpathian foredeep. The former are distributed in the Carpathian region through the Inner and Outer zones at the platform and the latter are situated only in the Inner and Outer zones of the foredeep [29], associated with clayey rocks. The organic matter included in them transformed into petroliferous hydrocarbons due to the catalytic action of the constituent clay minerals. The latter also influenced the composition and properties of the hydrocarbons during their migration into reservoirs. Clayey rocks situated along the paths of movement of hydrocarbons (HC) not only assisted in the formation of oil pools, but also protected them from dissipation.

Still, until the commercial oil-fields were struck in the clayey reservoirs of the Bazhenovian formation of western Siberia, the question of clayey rocks as oil reservoirs remained unanswered, although oil and gas accumulations intimately associated with clayey rocks have been exploited in the USA from the beginning of the twentieth century. Today, oil and gas finds in clayey reservoirs are known in many parts of the world. The stratigraphic interval of their distribution is very wide, ranging in age from the Miocene to the Devonian.

Oil in clayey reservoirs was first encountered in California in the Santa Maria basin (USA). Later, there and in other basins of various countries, small, medium and large oil and gas deposits were struck (the USA, Italy, Africa and so forth).

In the USSR, oil deposits related to clayey rocks were struck in the Bazhenovian formation of western Siberia towards the close of the 1960s. This aroused interest amongst investigators to look at similar deposits as potential sources for new oil and natural gas reserves. Investigations initially centred on grouping clayey reservoirs according to the category of deposits in the Greater Salym and western Siberia [28], where commercial flows of oil were encountered from such rock layers. The results of these investigations were highlighted in a large number of publications, which examined the geochemical characteristics of the basin of the sedimentation of the Bazhenovian period, its depth, rate of accumulation of sedimentary material, palaeontological remains, composition and degree of katagenetic change in mineral and organic constituents, the properties of rocks as reservoirs, conditions of formation of their potential as reservoirs etc. However, in spite of the considerable volume of such investigations, many problems have yet to find specific solutions even today.

The absence of a comprehensive work on the well-known and potential clayey reservoirs emphasises the need for utilising such studies in the prognosis of oil-bearing rocks of the Bazhenovian formation throughout the territory of western Siberia and also for finding similar deposits in other regions of the USSR. This book attempts to fulfil this long-felt need. The data largely comprises the findings of investigations carried out by the author in the clayey reservoirs of the Bazhenovian formation of western Siberia, Domanikian horizon of the Volga-Ural region, Khadumian and Batalpashin formations of the lower Maikopian in eastern Cis-Caucasus and also oil pools where clayey rocks predominate as reservoir rocks: rocks of the Menilitovian formation of the Carpathians, Kuonamian formation of eastern Siberia and others. In presenting this material, data published by other investigators on different aspects of this problem have also been highlighted.

This book not only discusses the conditions of segregation of mineral and organic components of clayey reservoirs, and geochemical and structural-textural interaction of the constituents of the rocks, but also the formation of reservoir potential and the oil and gas accumulations in them, as well as reasons for the appearance of a high-pressure anomaly. Prognostic delineation of the territory of distribution of industrially potential clayey reservoirs constitutes the ultimate aim of these investigations.

To determine the role of organic matter in formulating the infiltration parameters of clayey reservoirs, the author conducted a series of experiments, predominantly on organic matter of a humic and sapropelic nature in contact with rock-forming minerals of clayey reservoirs. The resultant data emphasises the importance of revising the system of secondary methods and of intensifying studies concerned primarily with clayey reservoirs.

A comparison of the composition, structure and conditions of post-sedimentary development of clayey reservoirs with the history of the tectonic development of the regions in which such reservoirs are found, enabled certain conclusions regarding the characteristics of formation of deposits in clayey reservoirs and the time of their being filled with oil, and the formulation of norms for detecting oil in similar rocks in different regions.

Contents

1
Methods of Study of Clayey Reservoirs

1.1 Material Composition of Clayey Reservoirs and Methods of Study

Clayey reservoirs of oil and gas constitute complex natural systems, the components of which are represented by rock-forming and accessory minerals, organic matter (OM) and pore waters. The reservoir potential and other properties of these rocks are influenced by the geochemical interaction of adjacent components with each other in the process of lithogenesis. The scale of this interaction is determined by the textural pattern of the rock.

The polymineral and finely dispersed nature of the constituents of clayey reservoirs necessitates utilisation of complex methods for their investigation, of which the principal one is mineralogical (study of thin sections and liquid immersion study of grains under a polarising microscope, study of suspensions in an emission electron microscope and analysis of massive rock samples under a scanning electron microscope). This method involves a study of the nature of authigenic minerals in clayey reservoirs, the types and forms of distinguishable organic matter (OM) and the textural features of the rocks. It is necessary to emphasise that the pore space of rocks–a textural element–changes concomitant with a change in their textural pattern, which determines the nature of the interactive behaviour of the components of the rocks and their spatial orientation.

Following M.S. Shvetsov and E.M. Segreev, the author distinguishes macro-, meso- and microtextures based on the scale of exposition of textures. In formulating infiltration properties of clayey reservoirs, meso- and micro- textures are essential. We notice, however, that in clayey reservoir rocks in which the principal mineral components are clay minerals (Bazhenovian formation of western Siberia, Menilitovian formation of the Carpathians, lower Maikop formation of eastern Stavropol' etc.), zones of linkage of microtextures constitute the major portion and zones of linkage of mesotextures are subordinate. In clayey reservoirs where carbonate minerals are the chief mineral constituents (Domanikian horizon of Ural-Povolzh', Kuonamian formation of western Siberia etc.), the zones of linkage at the mesolevel predominate and those of linkage on a microscale are subordinate.

Clayey reservoirs containing commercial reserves of oil and gas exhibit laminated mesotextures and axial microtextures. A lithological diversity with massive mesotextures and unoriented microtextures transposed on the reservoir lithic layers, usually does not contain commercial reserves of oil. It was observed that whereas the reservoir characteristics of rocks with disorderly (massive) mesotextures and unoriented microtextures are the same in all directions, rocks with laminated mesotextures and axial microtextures possess anisotropic reservoir properties.

Differences in the infiltration characteristics of rocks with various textures arose from the development of zones of textural banding of microblocks of clay minerals (acting as a single crystal), microlenses and bands of silty material, concretions of carbonate minerals and organic matter, which were not confined to the pores but formed bands of increased permeability.

During the vertical migration of hydrocarbons (HC) through the pores of rocks with laminated mesotextures and axial and clustered microtextures, the weakened zones became exposed. This development took place primarily perpendicular to the bedding plane or nearly so, although the patch of weakened parts situated along the bedding or close to it opened up simultaneously. During lateral migration the weakened zones developed mainly along the layering of the rocks. However, the oil and gas accumulations in clayey reservoirs indicate that characteristically the formations are typically heterogeneously distributed in space and, perhaps, that this happened primarily during vertical migration.

Mesotextures can be differentiated into parallel-banded, petal-like, lenticular-banded types of layers and lenses of organic matter, terrigenous or carbonate minerals distributed in the main rock mass. Microtextures close to axial are rarely clustered or oblique. Clustered microtextures formed in rocks by the distribution of parts with organic matter, recrystallised silica and, similarly, interrelated microblocks and microaggregates of clay and carbonate minerals. Enveloping microtextures appeared when centres of stress were present, which may have been a grain of quartz, crystal of calcite, concretion of silica or patches of carbon-fixed organic matter (OM) around which clayey and fine-grained carbonate minerals exhibit 'kinking'.

The textural framework of rocks is closely related to their material composition. Textural heterogeneity is due to the polycomponent nature of clayey reservoirs. The boundaries of textures of different types constitute zones of weakness. For clayey reservoirs, these weak zones (Fig. 1) formed the major portion and chief channels of migration of HC both during the genesis and accumulation of oil pools.

The basis of textures and consequently of the pore system of clayey reservoirs is rooted in sedimentation (primary textures and pores) and is mainly contained in diagenesis and katagenesis. The post-sedimentation history of existing rocks reveals traces of sedimentary structures and pores which have retained their identity in spite of post-sedimentation changes.

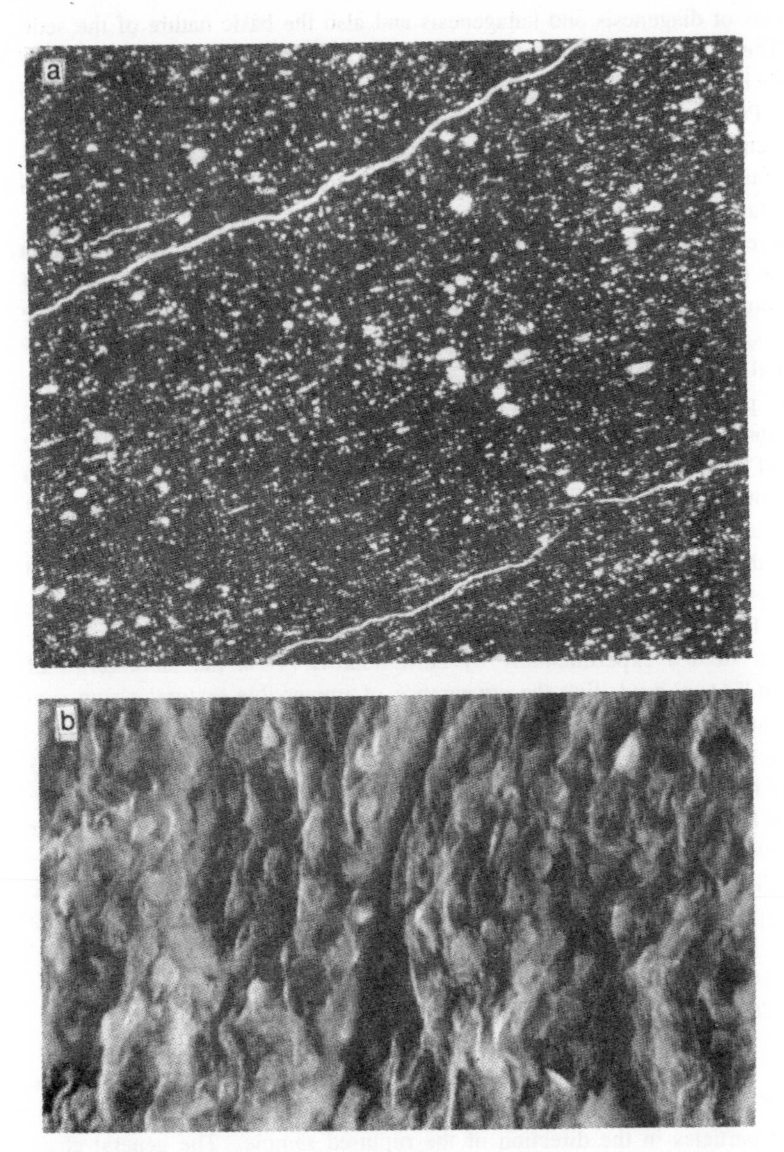

Fig. 1. Weakened zones in clayey reservoirs at the meso- (a) and microlevel (b).

a—Salym area, well 49, depth 2821.1–2.835.8 m; magnification 15, Nichols II; b—Zhurav area, well 62, depth 2153 m; magnification 1000.

The degree of preservation of primary attributes in rocks right from the time of deposition depends upon the physico-chemical and temperature-pressure conditions of diagenesis and katagenesis and also the basic nature of the sediment. The retention of sedimentation parameters in rocks is an expression of the physico-chemical laws of formation of sedimentary rocks, as established in 1933 by L.V. Pustovalov and developed by A.N. Dmitrievskii [13], which emphasise the tendency of sedimentary bodies during post-sedimentation alteration to preserve their primary features manifesting the basic controls on transportation and sedimentation.

Clayey rocks possess the maximum capability for retaining sedimentation indicator characteristics. Their low permeability reduces the energy flow of post-sedimentation processes. This property is particularly conspicuous in those clayey rocks whose specific composition and texture make them ideal reservoirs of oil and gas.

The property of clayey reservoirs depends on the presence of diagenetic silicification and a significant quantity of sedimentation of OM throughout the period of existence of specific conditions leading to the appearance in them of siliceous and organic pockets in the microblocks and microaggregates of clayey and carbonate minerals (Fig. 2). Hence it is thought that the katagenetic stage in the formation of clayey reservoirs exerted no influence on the mineral composition, textures and pore system as such.

The formation of textural peculiarities in clayey formations during diagenesis was studied experimentally by A.M. Tsareva [4] at the laboratories of VSEGINGEO by modelling the appearance of irreversible textural changes in water-saturated clayey samples during uniaxial and triaxial compression. Investigations were carried out on clays of undisturbed nature containing the clay minerals hydromica and kaolinite and the non-clayey silica, feldspar and chlorite. The natural moisture in the Iol'dievian clays exceeded the plastic limits by 8% on average. Experiments were done on the stabilometer S-1.

Specimens subjected to uniaxial compression until rupture showed a large plastic type of rupture and A.M. Tsareva observed two kinds of textural changes in them. Textural changes of the first type, resulting from the action of principal (compressive) stresses, are of particular interest to us because they predominate in diagenesis. These changes are related to orderliness in spatial orientation of particles because of the relative action of these stresses and result in better axial textures. Textural changes of the second type, formed by the action of tangential stresses, led to rupture textures and restructuring of orientation of particles in the direction of the ruptured sample. The general character of textural changes involving principal stresses is the same in the case of rupture with or without displacement. Differences appear only in changes in the nature of textures in the same zone of rupture facilitated by tangential stresses.

In triaxial compression tests, samples were characterised by a plastic type of deformation but showed no zone of rupture. Spatial orientation of the compo-

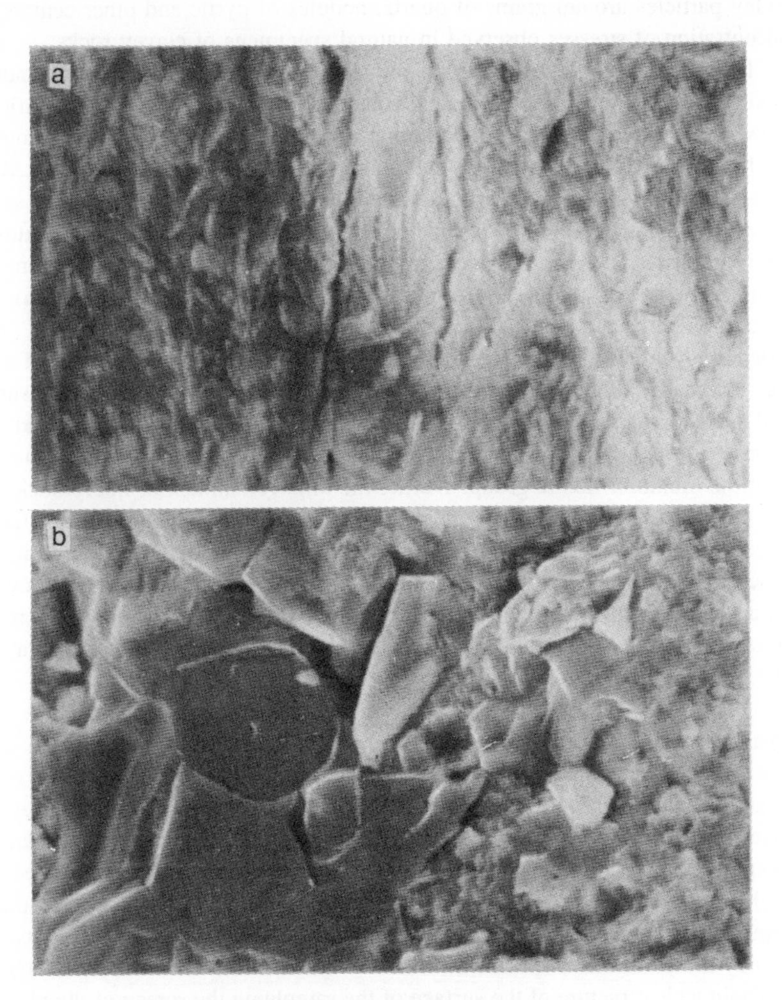

Fig. 2. Silica pockets in clayey fractions (mag. 1000)

a—Zhurav area, well 62, depth 2104–2112 m;
b—Skhodnits area, well 17, depth 4912 m–4920 m.

nents was well represented in all the comparatively equigranular samples. During uniaxial compression the maximum disposition of preferred orientation of the particles was seen in the zone adjacent to the central part of the specimen, with gradual deterioration towards the edge. In the case of maximum deformation, when the clayey particles lay between grains of quartz and feldspar, concentration of stresses was observed. This feature explains the process of reorientation of clay particles around grains of quartz, nodules of pyrite and other centres of concentration of stresses observed in natural specimens of clayey rocks.

The influence of katagenesis on the texture of clayey rocks is best understood from the results of experiments conducted by us on the action of various stresses on samples of Oglanlin bentonite. The natural moisture of the samples was 12.5%. Analysis of original and deformed samples under a stress of 50 MPa and a temperature of 20°C clearly indicated that the particles of the original sample remained flat, bringing out their fine sculpture, ordering of primary textures was promoted and the microblocks were localised. These changes were more prominently seen in samples deformed under a stress of 200 MPa (Fig. 3a). In a specimen deformed under a load of 50 MPa and a temperature of 200°C, the tendency towards distinct decrease in microblocks (Fig. 3b) and ordering of textures increased, although it is not possible for montmorillonite clays to assume ideal axial textures, which favour a large amount of structural water around the cations (neutralised lattice nodes), which helps in the high dispersion of minerals of the montmorillonite group. During the impoverishment of microblocks of clay minerals, belts of increased permeability are formed, thereby substantially increasing the permeability of clay minerals in general and clayey reservoirs in particular. Thus a study of the behaviour of clay minerals in katagenesis showed that an increase in pressure did not change the type of textures but simply reinforced them to enable localisation of microblocks. An increase in temperature led to more energy dissipation along the different textural units.

Utilisation of optical (OEM) and scanning (SEM) electron microscopes has considerably widened the spectrum of mineral investigations of such finely dispersed formations as clay minerals. The simplest and most extensively used method of investigation of clay minerals with OEM is that of suspensions. It involves estimation of form, sizes, characteristics of grain boundaries and relative thickness of the clay mineral particles, i.e., their morphological peculiarities.

Recourse to the SEM method in preference to the OEM is largely determined during morphological studies by the large depth of focus, whereby it is possible to study the unfractured specimens with highly varied relief in different targets. To obtain a clear picture of the surface of the sample on the screen of the picture tube, secondary electrons are partly directed to the surface of the specimen with an energy sufficient to overcome their potential barrier [10].

To protect the surface of the sample from external charges and to obtain a picture of high quality in a vacuum, a thin layer (20–30 nm) of gold is coated over the sample during rotation, which protects the emulsion and prevents the sample from oxidation in air. With the help of a scanning electron microscope (SEM), a wide range of geological problems are solved, amongst which the most important in the study of clayey reservoirs include major forms of silica secretions, character of weakened zones at the microlevel, spatial distribution of authigenic minerals, structure of pore space etc. A large number of users

Fig. 3. Separated weakened zones in montmorillonite clay at 200 MPa and 20°C (a) and 50 MPa and 200°C (b) magnification 1200.

of the SEM facility increasingly feel the need for easy transfer of data from the optical microscope by means of drawings on the screen of the picture tube. The effectiveness of the SEM is increased when used in combination with the OEM. Then, the advantages of both methods supplement and complement each other.

To determine the clayey composition of reservoir rocks and also the type of microtextures, the *X-ray diffraction method* is used. A photograph of the samples can be obtained with the diffractometer DRON-2. The oriented samples are labelled with the following regimal data: C and K_α-radiation, voltage of 30 kV, current 12 mA, limits of measurements 1000 impulses/sec.

The absence of sharp reflections in the diffractograms of oil-saturated clayey rocks, particularly those from reservoir horizons in the Bazhenovian formation necessitated resorting to some method of preliminary processing of the samples to obtain improved quality of reflections.[1] Initially, the specimens were subjected to ultrasonic vibrations for 3 min and later for 15 min. The sharpness of the reflections registered no significant improvement in either ultrasonic experiment. To obtain an increased integral intensity of the basic reflections was thus practically impossible. The interaction with chloroform also did not change the nature of the first basal reflection of the hydromica (1 nm), which remained only wide and not high for the most part.

Samples were then processed according to the method followed at the Western Siberian Institute NIGNI. This method involves processing of the specimen for 10 days in a 15% solution of hydrogen peroxide and later washing in a 5% solution of oxalic acid for 6 hours. Diffractograms of the samples were recorded both before and after the washings and processing. It was found that the reflections of hydromica were not changed by this procedure and only those of feldspars remained conspicuous. Similar processing of rock specimens from the layered Bazhenovian formation led to a conspicuous increase in integral intensity of the first basal reflections of the associated hydromicas.

The results obtained from variously processed rock samples support the observation that hydromica components of clayey reservoirs favour the presence of a hydrophobic envelope in the adjacent clay minerals and are formed by the adsorption of organic matter (OM) and silica. They do not disintegrate during the preparation of samples for analysis. We believe that these envelopes need not disintegrate, because after their formation the property of the rocks disappears, so that the oil-saturated clayey reservoirs differ from the overlying and underlying formations and, further, they also differ in values of electrical conductivity (EC) and gamma-logging (GL) characteristics.

[1] X-ray investigations of the specimens were carried out by I.F. Metlova.

In addition to the X-ray diagnostics of the clayey constituents of the reservoirs, their polytypes were studied by the *method of electronography*.[2] Electronograms were obtained from conical textures (Lα = 60°) on the electronograph ER-100 during acceleration of stress to 100 kV.

The *thermographic method* of study of rocks is widely used in mineralogical investigations. The polymineral character of clayey rocks, however, makes it difficult to interpret the thermogram–not only the total thermal effects of the constituent minerals, but also those related to the appearance of new phases during the thermal interaction of substances comprising the rock under study. The latter situation is particularly conspicuous when considerable amounts of dispersed OM, characteristic of clayey reservoir rocks, are present in the samples.

The author adopted a thermoanalytical method for obtaining information on the composition of the dispersed OM and its regenerative potential. Concentrates of OM were obtained by separating the inorganic part from the rocks. Carbonates were disintegrated by treating the rocks with 10% hydrochloric acid and heating them for 30 min and the silicates were removed by treatment with 45% hydrofluoric acid for 16 hrs. A preliminary separation of soluble OM from the rocks was achieved by chloroform and spirit benzol extraction in a Sokslet apparatus.[3]

After acid disintegration of the inorganic part of the rock, the OM was concentrated by means of fractionation in heavy liquid (solution of cadmium iodide and potassium iodide with sp. gr. of 2.1). Processing of the OM was achieved by degreasing at a maximum temperature of 80°C for 3 hrs. An interpretation of the results of the thermogravimetric investigation of OM concentrates was carried out by N.M. Kas'yanova based on standard thermoanalytical characteristics of the dispersed OM (obtained from Brazilian investigators [48]), which reflects on the one hand the composition of OM, and on the other the degree of its remobilisation. Two stages of decomposition can be determined with the assistance of these standards–that characterising low temperature volatiles and the other, high temperature non-volatile fractions (temperature of reactions corresponds to 330 and 460°C).

The following serve as criteria for distinguishing types of OM according to thermoanalytical characteristics: 1) relative size of the amplitude of exothermal effects, indicating the degree of mobilisation; 2) morphology of high temperature peak, also characterising the degree of mobilisation; and 3) morphology of peak during 460°C, during low mobilisation of OM is revealed by corresponding inert, subcolloidal and cellular-vegetal types of OM. Based on differential thermal analysis (DTA) curves, four distinct types of OM are distinguishable: inert,

[2]Determinations were carried out by V.F. Chukrova.

[3]Extraction from the rocks was conducted by M.F. Lobanova.

subcolloidal, cellular-vegetal and mixed subcolloidal and cellular-vegetal.

Special instruments were required to ascertain the textural framework at different levels of study. The mesotextures were studied in thin sections under a polarising microscope and the various types of laminated mesotextures determined for clayey reservoirs. The microtextural characteristics of rocks were examined in thin sections according to the character of the aggregate extinction using special X-ray photography.

The qualitative description of microtextures according to the character of the aggregate extinction was supplemented by quantitative assessment using the coefficient of orientation $C = 1 - I_{min}/I_{max}$ suggested by V.I. Muravev in 1961, and the coefficient of relative ordering $V = S_{bl}/S$ suggested by V.M. Shibakova in 1963, where I_{min} and I_{max} = light rays during extinction and clear field of vision, S_{bl} = area of the thin section occupied by microblocks, and S = area of the entire thin section. During the absence of orientation of grains, the coefficient C is close to zero; the higher the degree of orientation of the grains, the closer C is to 100%. Determination of the light rays was carried out over the photoelectric microattachment FME-1. When C = 100%, the whole rock is composed of microblocks with an axial orientation.

A semi-quantitative estimate of the spatial distribution of the mineral grains at the microlevel was accomplished through special X-ray photographs. To obtain a complete picture of the spatial orientation of single crystals it was necessary to take a series of X-ray photographs for all rotations of the sample through 90° towards the direction of the X-ray source. If the direction of axis of orientation is known, then during axial microtextures it is possible to arrange for two photographs–one in which the axis is oriented parallel and the other in which the axis is oriented perpendicular to the X-rays. The photographic unit and the method of interpretation of the data obtained have been detailed by G.K. Bondarik [4].

1.2 Study of Structure of Pore Space of Clayey Reservoirs Using the Electronic Computer Quantimet-720 and the Scanning Electron Microscope

To study the pore space structure in clayey rocks, the method worked out by P.A. Konysheva and A.P. Roznikova in 1976 was used. According to this method, pores are effectively highlighted in the non-printed photographs from a scanning electron microscope. A photograph obtained by this method was fed into the electronic counter Quantimet-720, which basically works on the principle of scanning the figure obtained by means of a microscope possessing both transparent and reflected light or an epidiascope in which photographs, negatives, figures or polished samples can be placed. Scanning the plane of the figure is done by standard television methods. By means of a cathode ray tube (vidicon or plumbicon), an image is obtained for a large quantity of elements, which are grouped into 720 scan lines and processed into a series of electronic signals with an amplitude proportional to the sharpness of the object. A further signal reaches the discriminator block (detector) and the signals corresponding to the object under study are isolated from the general information.

The detector possesses three threshold regulators designed to measure four components which differ from one another in their transmittance (contrast) or reflection characteristics. The detectable signal consists of a series of right-angle impulses, the heights of which are constant and the lengths correspond to the lengths of the chords crossed by the contours of the detected particles.

A parallel drawing of the figure projected from the microscope or epidiascope onto the monitor screen enables observation and accurate measurement of the required particles. An accessory construction apparatus facilitates simultaneous projection of the natural and the drawn figure on the screen, enabling the computation of a variety of functions.

The detected signal from the chosen particles is directed to the block of the standard computer where different operations are executed to obtain the necessary parameters.

Under a magnification of 1200 and 3600 eight electron photographs were miscalculated–five along the layering of the rocks and three perpendicular to it. The latter three photographs were scanned in two directions, across and perpendicular. The following parameters of the pore space of the rocks were then determined: pore area, average diameter of chords of pores, specific surface (number of scan lines crossing the contours of pores is fixed at a unique interval, i.e., 1 mm), distribution of pores according to their size, and orientation of pores and fissures. Fig. 4 depicts an SEM photograph of a reservoir rock from the Bazhenovian formation in the Salym oil-field (low yield). The quantification results were obtained by P.A. Konysheva and A.P. Roznikova [22] on a Quantimet-720.

In these SEM photographs the interrelation of the rock components, the character of the pores and the pore canals connecting the individual pores as varied forms of a branching pattern are readily discernible. In the graphic display the distribution of pores is well delineated so that under a magnification of 1200 the maximum quantity of pores is observed at an interval of 0.22–0.43 μm but under a magnification of 3600 the interval is seen to be 0.05–0.15 μm. In the background of small sinuous pores, pores of larger dimensions, say 3–5 μm, are observable.

To obtain comparative data on the parameters of pore space of the rocks of the Bazhenovian formation in open high-yielding wells, rock samples from well no. 32 of this deposit were investigated according to the method described above. The results, generally depicted in the form of differential and integral

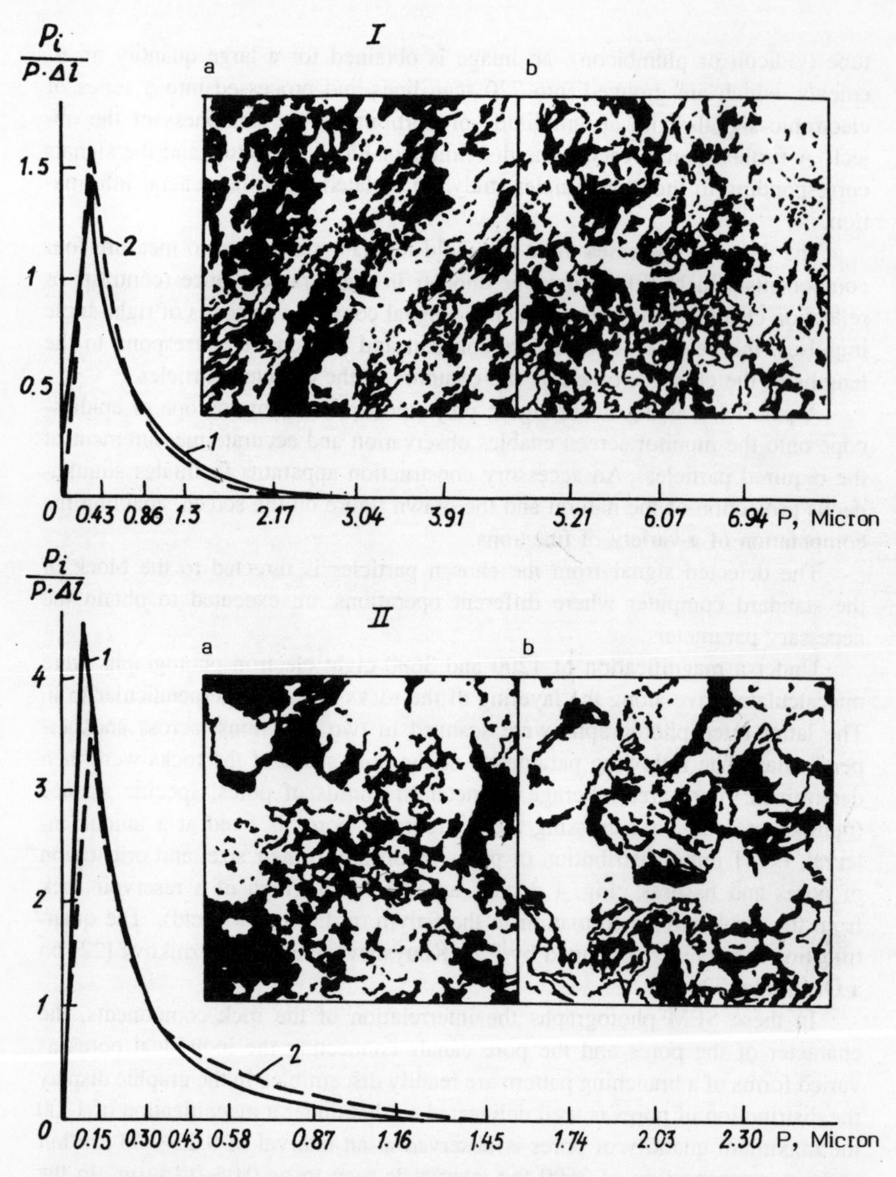

Fig. 4. Photograph from scanning electron microscope (I and II correspond to magnifications of 1200 and 3600 respectively;

a and b—different fields of one photograph) and curves of distribution of sizes of chords of pores n (1–to photo a and 2–to photo b).

P—number of chords of pores of given size;

P_i—total number of pores;

Δl—interval of measurements of chords of pores in μm.

curves, are shown in Fig. 5. Comparative data on the parameters of pore space in the clays of the Bazhenovian formation from wells of varying yields are presented in Table 1.

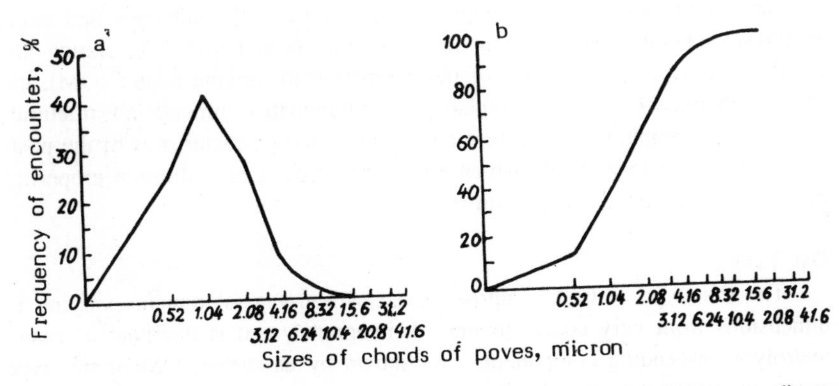

Fig. 5. Differential (a) and integral (b) curves of distribution of chords of pores according to sizes (magnification 1200). Salym area, well 32, depth 2775.31–2779.97 m.

Table 1: Qualitative indicators of microporosity of clays of the Bazhenovian formation (under magnification 1200)

Zone	Sizes of pores		
	Mean	Predominant	Maximal
Low yield	0.70–0.75	0.50	0.87
Maximal yield	2.20–2.30	1.04	2.40

The adoption of this methodology demands an explanation of the role of pore space in the formation of commercially productive clayey reservoirs. As seen from the data shown in Table 1, high yields are characteristic of rocks with pores three times larger than those present in rocks of poor yields. Lesser size or number of pores in rocks of high-yielding wells (maximum porosity 8.5%) do not guarantee such outputs.

The methodology discussed above was adopted by the author for assessment of the extension of the weakened zones at meso- and microlevels. To determine the length of the linked zones of mesotextures, photographs of thin sections were used and for linked zones of microtextures, photographs of SEM.

1.3 Determination of the Influence of Organic Matter on the Capacity Properties of Clayey Reservoirs

Many years of mineralogical and petrological investigations on OM from deposits associated with oil-gas-pools of both young and ancient platforms carried out by the author during 1970–1980 led him to establish that irrespective of the age, grain size and genesis, there are three types of OM present, which vary in quantity and can be distinguished according to the influence they exert on physico-chemical environment in the sediments and rocks. Changing under the action of the processes of transformation of organic matter (OM), the physico-chemical conditions stimulate post-sedimentation mineral and structural-textural transformations in the formations. Only the characteristics of the post-sedimentation history of OM which typify the capacity and filtration properties of clayey reservoirs are dealt with below.

OM, TYPE I

The first type of OM comprises fixed carbon-vegetal remains varying in dimensions from very coarse to fine, disintegrated, and is deprived of easily hydrolysed ascending components. As shown by us earlier, OM of this type exerts no influence on the physico-chemical environment and does not stimulate new formation of minerals around its parts. It only serves as part of the matrix in which authigenic minerals of carbon, titanium and iron are formed. The formation of clayey reservoirs is characteristic of marine basin sedimentation far removed from eroded dry land, which is conducive to the supply of finely dispersed, washed-out material, including OM of the first type, and invariably only pyrite is formed in the last stage. This is the main source of pyrite in clayey reservoirs. Often finer particles of pyritised OM of the first type get grouped into unique colonies, which develop as concentrators of stresses oriented around thin particles of clay and carbonate minerals. Zones of contacts occur between these and individual globules and with weak colonies of pyritised OM. In clayey reservoirs they provide additional space in and through which the hydrocarbons can migrate.

OM, TYPE II

The second type of OM comprises vegetal remains with varying contents of hydrolysed components and also fragments of spores, clots of OM of humic or sapropelic nature, which also secrete hydrolysed components during transformation. The latter penetrate the rocks formed around the material OM aureole, the size of which is controlled by free space between the particles of rock and the sorption volume of the surrounding minerals. In addition to the hydrolysed components from OM of the second type, there are soluble forms of titanium and iron. At some distance from the OM body crystals of siderite, anatase and pyrite form which, however, are not related to OM at all. When crystals of the said minerals colonise adjacent to the clayey components of the rocks, they then serve as centres of concentration of stress and thus favour the reorientation of the latter. Zones of weakness are produced between the newly formed minerals and the main clay mass. The essential property of OM of the second type is its participation in providing the characteristic textures at the mesolevel typical of clayey reservoirs, primarily parallel to the plane of bedding of the rocks, resulting in the creation of anisotropism of infiltration parameters. In the process of late diagenesis and early katagenesis in the molecules of the humus part of OM of the second type, oxygen links break up, releasing carbon dioxide which dissolves potential series of carbonate minerals, thus increasing the capacity of clayey reservoirs.

OM, TYPE III

The third type of OM occurs in any geological material as true or colloidal solutions, the sizes of its components not exceeding those of exchangeable positions of the clay minerals ($\sim 1\ nm^2$).

A series of experiments on the transformations of OM of different classes in clay minerals of various composition were conducted to establish the mechanism of interaction between clay minerals and organic matter. In the first series various organic materials–oleic acid, blue-green algae, antarctic zooplankton, brown coal of low stage katagenesis (not tested for interaction of heat)–reacted with only one clay mineral–montmorillonite. In the second series only one kind of organic matter–oleic acid–reacted with various clay minerals (kaolinite, hydromica, montmorillonite). In the third series oleic acid reacted with collomorphic silica and carbonate calcite (Iceland spar).

Several classes of organic compounds were obtained as a result of these experiments: hydrocarbons (HC) of varying structure and molecular mass, neutral oxygen co-ordination (including ketones) products, irreversibly adsorbed minerals and mixture of acidic products. In the experiments conducted without water, the ratio of isonormal hydrocarbons to normal hydrocarbons was much higher than in the experiments with water. A study of the minerals before and after the experiments, optically and by X-ray, thermographic and electron microscopic methods, showed that the principle of interaction, i.e., the basic mechanism of interaction of OM with minerals of the sedimentary rocks is sorption of organic ions or radicals on the surface of the minerals and particularly their parts since here, because of defects in structure, a deficit of positive charges is created. Hence, there is an indirect indication of the property of sorption by the minerals and that means transformation of OM, i.e., an increase in its exchange capacity [45].

The property of minerals to adsorb organic ions or radicals is influenced by certain factors, amongst which the surficial strengths directly related to proportional dependence on the specific surface of the minerals or, in other words, their dispersion, plays a vital role. Increase in the exchange capacity of minerals is also related to the isomorphous replacement of major cations in the lattice by cations of lower valency and also to the hydration of the surface, i.e., ratio of number of oxygen to hydroxyl ions in them.

The transformation of OM takes place at the expense of the internal energy contained in it. However, this energy is not put to use without the aid of outside factors. One of the stimulating conditions which drives this energy in OM to action is sorption by its minerals, which adsorb organic ions or radicals, lower the energy of their activation and thus assist in their transformation. Between the sorption of organic ions and the minerals a distinct link is established whose strength varies within wide limits, depending on the size of the organic ions. The larger the latter, the stronger their sorption by the surfaces of the minerals, because the organic ions, besides their Coulomb-bonds interacting with differently charged particles, are held by additional Van-der-Waal bonds. The larger the latter, the larger the size of the organic ion or radical.

Because of their typically high sorption and the presence of exchange centres, all clay minerals possess the property of adsorbing organic ions and radicals. Amongst the clay minerals, the maximum exchange capacity is exhibited by minerals of the montmorillonite group and the minimum by those of the kaolinite group. Montmorillonite, with an exchange capacity of 68.5 mg- equiv/100 g, and prosyanovian kaolinite, with an exchange capacity of 6.7 mg- equiv/100 g, were used in the experiments.

Based on the sorption characteristic, the minerals of the hydromica group are intermediate between the minerals of the montmorillonite and kaolinite groups. To increase their sorption capacity, both isomorphic substitution in the lattice and increase in the degree of dispersion led to the formation of uncompensated co-ordinations. Special significance is attached to the minerals of the hydromica group, which exhibit substitution in the tetrahedral sheets of tetravalent silicon and trivalent aluminium, whose transition in the four-fold co-ordination corresponds to the formation of uncompensated bonds and increase in exchange capacity.

The organic matter of the third type and the hydrolysed components of OM of the second type are adsorbed by the clay minerals according to the rules of chemical (ionic) sorption, which is accompanied by the escape from the active centres of minerals of exchangeable ions together with the water molecules surrounding them and the formation of hydrophobic coatings on their surface. And, as the clay minerals possess the property of forming microblocks and microaggregates which behave as one single crystal, not just the surface of one mineral only becomes hydrophobic; a general hydrophobic film forms at the contacts of different textural zones. This property of sorption of organic ions by the surface of clayey microblocks leads to the formation of typical textures with organic matter (OM), which are extremely important for the formation of a large capacity in clayey reservoirs, as the boundaries of microtextures with OM are far weaker than zones of contacts of various types of microtextures without OM.

Another group of rock-forming minerals in clayey reservoirs, which possess high sorption capacity, are the minerals of the silica group present in the rocks as collomorphic bodies at different stages of recrystallisation. As is well known, the structure of the minerals of the silica group presents a framework in which all the tetrahedra are connected at the top so that each silicon ion is surrounded by four oxygen ions and each oxygen ion is shared by two silicon ions. Collomorphic silica is characterised by disorderly spatial orientation of these tetrahedra. The sorption property of the silica group of minerals is determined by partial substitution of silicon in the tetrahedra by aluminium, which results in a deficit of positive charges being compensated by exchangeable ions, of which the most important are those of hydrogen. The minerals of the silica group are also characterised by the property of physical sorption by organic molecules, which are held on the surface of minerals by Van-der-Waal bonds. The greater the bonds, the larger the organic molecules.

Because of the fact that collomorphic silica facilitates not only chemical, but also physical sorption of OM, the absorption capacity of silica before and after treatment with oleic acid dwindled significantly more than the exchange capacity of montmorillonite under a similar experimental set-up. The residual content of organic carbon (C_{org}) in the colloidal silica was higher than twice that in montmorillonite (Table 2).

Table 2: Absorption capacity and C_{org} of minerals before and after treatment with oleic acid

Minerals	Absorption capacity mg-equiv/100 g		C_{org},%	
	Before experiment	After experiment	Before experiment	After experiment
Montmorillonite	68.5	53.9	0.24	5.46
Collomorphic silica	47.8	24.2	0.30	11.50

In the four-component clayey reservoirs of the Domanikian type, carbonate minerals such as calcite and dolomite constitute the dominant rock-forming minerals. The latter plays a subordinate role compared to calcite. The sorption capacity of both is mainly due to surface forces, which increase with greater dispersion of the minerals. Due to the incompletely compensated bonding strengths of atoms situated on the surface, the particles are subjected to comminution, or the minerals appear with an incomplete crystal structure. As demonstrated by R. Grim and M.F. Vikulova, the broken bonds developed *in situ* might form the sites of sorption or organic cations. In the experiment on the transformation of OM in carbonate minerals, the Iceland spar used was prepared, similar to other minerals, in the form of granules 3–4 mm in size. In the experimental process, in spite of their low sorption capacity, the carbonate minerals assisted in the transformation of OM, which proceeded along one direction, whereas the process was less intense in the clay minerals. The smaller the fragments of the carbonate rocks, the greater was their sorption of OM. It is, however, pertinent to point out that the scale of chemical sorption of OM by carbonate minerals is not high. The residual OM of the carbonate part of the rocks consists of components which have been adsorbed according to the laws of physical and chemical sorption.

The type of OM is reflected in the intensity and direction of its transformation. This is particularly important in deciding the property of the newly formed substances desorbed by the mineral surfaces. The quantity of residual OM depends upon the type of OM, i.e., the composition of its components, the sizes of which facilitate their sorption by minerals and sorption capacity of the minerals, and the OM cannot be removed by solvent extraction. Thus, during the interaction of montmorillonite with oleic acid (third type OM) the residual content C_{org} was nearly 5.46%, with brown coal (second type) 3%, brown coal in a mixture of oleic acid in the proportion 2:1 (second and third type OM) 3.69%, zooplankton (second type OM) 2.73% and blue-green algae (third type OM) 0.08%.

In one and the same type of OM the quantity of residual OM is proportional to the sorption capacity of minerals, that is, the more intense the activity of minerals in transforming OM and HC of various classes, the more difficult desorption of residual organic matter with their active centres becomes. Thus, during the interaction of clay minerals with oleic acid the residual content OM was 5.46% in montmorillonite, 1.5% in calcite and 0.96% in hydromica.

An idea of the quantity of irreversibly adsorbed OM can be obtained from the heating curve of rocks. The irreversible sorption of OM in the heating curve of montmorillonite after treatment with oleic acid produces a sharp exothermal kink at the temperature interval of 800–865°C, deforming the thermal curve and completely subordinating the third endothermal effect of montmorillonite. A not so intense, very wide exothermal effect is also observed at the interval 250–550°C, which appears as though divided into two parts by a weak endothermal effect. This exothermal effect is related to the oxidation of OM of the oil series not released from the rocks by spirit-benzol extraction. An endothermal reaction, maximum at a temperature of 480°C, favoured dissociation of some part of the OM irreversibly adsorbed by the surfaces of clayey particles, according to the data provided by J. Jordan.

In the differential thermal curve of kaolinite, after similar treatment, a sharp exothermal peak was observed at the temperature range of 730–820°C, favouring improverishment of hydrogen of the heavy residue, closely associated with the surface of the mineral by its burning down. This exothermal bend was singularly unique, distinct from the thermogram of the ignited original kaolinite sample.

The curve of heating of hydromica, after treatment, differed from the thermogram of the original sample by a small exothermal kink at the temperature range 640–760°C. Here, too, as in the thermograms of montmorillonite and kaolinite, there were links with OM irreversibly sorbed by the hydromica. Similarly, experimental data point out that the less the irreversibly adsorbed OM in the rock, far less will be the temperature at which the process of its burning out commences. The temperature ranges of minerals are 800–865, 730–820 and 640–760°C.

The presence of irreversibly adsorbed OM, which makes the surface of microblocks of minerals hydrophobic, in fact determines the reservoir potential of these rocks and provides conditions without which they could not contain and yield hydrocarbons, thereby attaining the status of reservoirs.

2
Clayey Reservoirs of Foreign Oil and Gas Basins*

2.1 Basins of the USA

SANTA MARIA BASIN

The most significant oil find in clayey reservoirs was first struck at Santa Maria basin in California. The oil-field covers Orcatt, Lompoc, West Cat Canyon, Kasmalia, Gato Ridge, Santa Maria Valley, Zaka Creek and other areas. Siliceous shaly clays of the Monterey formation (Upper Miocene) constitute the reservoir rocks of the pools of this basin. The formation consists of three parts: the upper platy siliceous shaly clays with OM, the lower siliceous shaly clays wherein silica is substituted by calcite and a middle layer of siliceous limestones. The general thickness of the formation in different parts of the basin ranges from 0 to 750 m, the average thickness being around 450 m according to the estimates of M. Hubbert and D. Willis.

Oil is confined to the anticlinal structure of the block (covering Kasmalia, Orcatt, West Cat Canyon, Lompoc) or to the zones of occurrence of monoclines (Santa Maria Valley) (Fig. 6). The oil accumulation occurs in laminated diapiric, tectonic, lithological and stratigraphic traps. Typical early outputs varied from 64 m^3/day (Kasmalia) to 400 m^3/day (Santa Maria Valley). Annual exploitation of the reserves in the pools of Santa Maria Valley, Orcatt, West Cat Canyon and Lompoc has yielded 360 million m^3 of oil and gas from the siliceous shaly clays of Monterey [38].

SAN JOAQUIN BASIN

This basin is asymmetrical with steep tight western and broad gently dipping eastern flanks. The western flank is complicated by projections of shore ridges and anticlinal zones and depressions in the sedimentary formation of the basin. The eastern flank possesses a homoclinal structure. It is complicated by longitudinal faults of not much amplitude and by depressions connected with

---

*Several place names given in this chapter were not verifiable from the Russian transliteration—Language Editor.

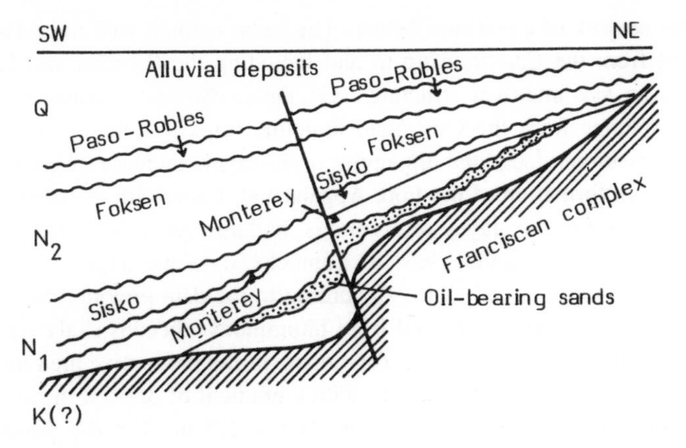

Fig. 6. Geological cross-section of the Santa Maria Valley formation (after Levorsen, 1970).

these faults.

A few oil reserves are known in the Upper Miocene bituminous, siliceous or calcareous shaly clays of the Mak-Lur series (the so-called brown shales). The formation comprise North Belridge, South Belridge, Elk hills, Miduei-Sunset and others situated along the tectonic fault on the south-west border of the San Joaquin valley. The oil-pools are located in the anticlinal folds.

Siliceous shaly bituminous clays of Cretaceous age are found in the Moreno formation (Kaoling region) in Oil City in the San Joaquin basin. The productive layers of this field lie at an average depth of 300 m and the reservoir thickness is about 300 m. The normal output is less than 16 m^3/day.

Within the limits of the San Joaquin basin, oil is also obtained from the Jurassic metamorphic and slaty shales of the basement. These are the formations of Edison and Mountain View. The depths of occurrence of productive shales are 1050 and 2040 m respectively. The formations form part of the large ledge of the basement–an uplifted block dissected by a series of faults (M. Hubbert, D. Willis). The Edison deposit is most interesting as it is situated in the eastern part of the San Joaquin basin, where the homoclinal Cenozoic and Quaternary rocks transgress the eroded layers constituting the Sierra Nevada ridge. Oil was struck in this formation in 1931 and in the course of 14 years two different layers of sandstones of the Cenozoic have been exploited. In 1945, one of the borewells was halted by the metamorphic rocks of the basement; after drilling through these oil flows up to 84.5 m^3/day were obtained. Initially, the outputs in the borewells was inconsistent but 25% of them produced more than 160 m^3/day of oil.

Fore-Appalachian Basin

This basin shows a relatively simple structure and principally consists of the buried framework of a platform bench. The basin is filled with thick Palaeozoic sediments from the Cambrian up to and including the Permian and invariably possesses an asymmetrical structure. On the north-west platformal ledge all the rocks plunge from the Cincinnati anticline (dips less than 1°) and on the south-east piedmont ledge the dip angles of the layers increase by some degrees.

The main formation of the Fore-Appalachian basin with commercially productive horizons in the clayey reservoirs is the large gas-field of Big Sandy and some allied gas-bearing units east of Kentucky state and adjacent parts of the states of West Virginia and Ohio. According to the data provided by K. Hunter and D. Young, the gas is obtained from bituminous siliceous shaly clays of the Ohio suite belonging to the Upper Devonian. In the productive area these rocks are homoclinally tilted south-easterly with a gradient of 3.5–5.8 m for 1 km. In this section the thickness increases from 105 to 225 m. The Big Sandy formation is a monocline (Fig. 7) with flexural folds and tapering rock layers. The oil-pool occurs in the layered, anticlinal lithological trap.

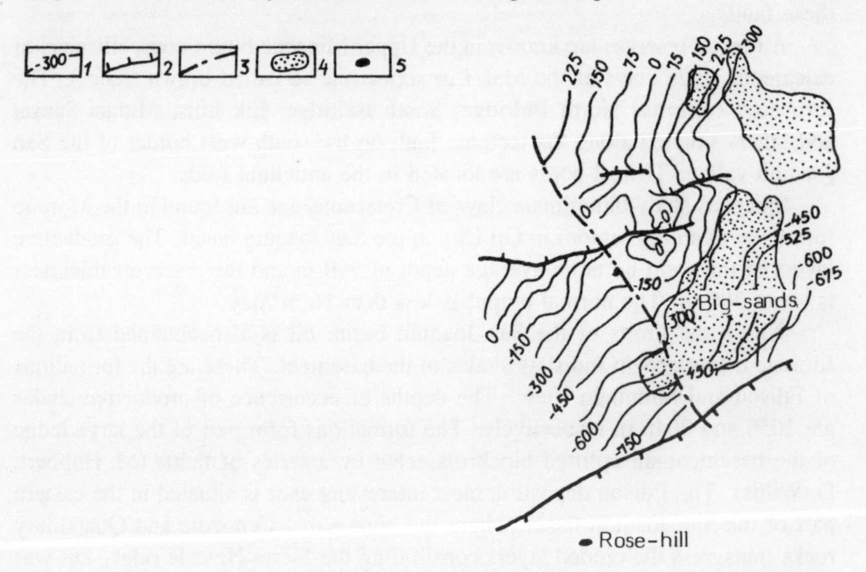

Fig. 7. Scheme of distribution of clayey reservoirs of the Fore-Appalachian oil and gas basins (from M. Hubert and D. Willis).

1—isohypses of the hanging walls of the Biri formation (in the eastern part) and Korniferus (in the western part) in metres; 2—fault, 3—boundary of distribution of the Biri formation of deposits; 4—area of gas exploitation; 5—oil-pool.

Gas was first obtained in 1921 from the bituminous siliceous shaly clays of the Ohio formation. According to the information provided by K. Hubert and D. Young, at the Big Sandy formation 3414 wells had been drilled as of 1951, of which only 207 (around 6%) yielded an industrial supply of gas, starting from an average yield of 29,700 m^3/day. In all the remaining wells commercial flows were obtained only after taking appropriate measures, mainly opening up the rock pile after an interval by means of 80% gelatinised nitroglycerin. As a result, 3207 wells showed an increase in production from 1700 to 8200 m^3/day; however, another 323 wells (about 10%) remained unproductive. In 1973, one millirad cubic metre of gas was obtained from the Big Sandy formation comprising bituminous, siliceous and shaly clays [38].

Unfortunately, there are no further data in the literature relating to these and other oil-fields in which the reservoirs are bituminous, siliceous, shaly clays and no information on the structural status of the wells that remained unproductive even after the adoption of artificial measures. It is highly probable that these wells are situated in such parts of the structures wherein the productive horizons perhaps have been destroyed by processes of erosion, or perhaps the wells were driven in parts removed from the faults, which play a major role in the formation of a useful exchange in clayey reservoirs. The latter case is due to the separate identity of differently textured parts of the rocks under the action of tectonic stresses.

The productive ranges of closely spaced wells driven into the rocks of the Ohio formation are not always situated at one and the same level and might be encountered throughout the entire thickness of the rocks. The distribution of pressure in the formations supports the consistency of the reservoirs in every given oil-field. For example, the pressure in closed wells comes down simultaneously along with the average pressure in the reservoir formation, which precludes the possible suggestion that gas is obtained from permeable sandy intercalations. M. Hubbert and D. Willis explain that there is only one characteristic, viz., permeability that is facilitated by fracturing.

In 1980, in the Fore-Appalachian basin in one of the drilled borewells (in the central part of New York state) within an interval of 150–290 m, gas was obtained to the extent of 28.3 thousand m^3/day from the middle Devonian brown shaly clay Hamilton formation [50].

UINTA BASIN

In the Uinta Basin (north-eastern Utah state and north-western Colorado state), oil is exploited from the bituminous, siliceous, shaly clays of Eocene age. The inner part of the Uinta basin is complexly built up of 4500 m thick Eocene lacustrine or fluvial deposits, of which nearly 1200 m are situated in the Green River oil-bearing shaly clay formations. The latter are exploited in the Roosevelt formations, which are confined to the anticlinal structural ledge, plunge north-westwards and lack a structural closure, and at Dushen, which is situated on the fault zone. The field of development of siliceous shaly clays is limited by the given basin and indirectly by the adjoining Green River basin.

Still, in spite of the limited distribution these clayey shales contain the largest potential reserves of oil in the USA.

The early yields and productive ranges of adjacent wells were highly variable and, in general, were typical of fractured reservoirs.

In the same basin the oil-producing fields of Rangeley are located on the northern border of the inner basin of the Douglas Creek uplift. The structure of the formation is a cupola (dimension, 12 km × 32 km) intersected by faults. The productive horizons comprise siliceous shaly clays of the Upper Cretaceous (Mankos formation) and the Upper Carboniferous (Morgan and Round Valley formations). The reservoirs are of the layered, anticlinal, lithologically screened type and in 1974, 2.7 million tonnes of oil were obtained from them [38].

DENVER BASIN

In the Denver Basin oil is obtained from the Florence and Cannon City formations of the Mankos formation composed of siliceous shaly clays situated in Cannon City eastwards from the front ridge of the Rocky Mountains (Colorado state).

The Florence formation is situated on the monoclinal tilt westwards of the wide eastern flank of a syncline. Dips in the formation range from 2.5° to 3° but eastwards register a sharp increase with layers dipping at 45°. There is no water in the reservoir. Cannon City formation is situated on the southern border of the basin and forms part of an anticlinal structural ridge plunging southwards. As in all other formations where clayey reservoirs are exploited, here, too, the productive ranges in the wells, situated in rows, are often located at different depths. The directions of interference of the borewells are oriented in the same manner as the trends of the predominant fractures observed in the outcrops. Wells exploited at shallow depths often suffer interference from wells of greater depths and vice versa.

In the Denver basin, the productive layers are the siliceous shaly clays of the Pirr (Upper Cretaceous) formation in the Florence-Cannon City oil-fields, situated on the southern border of the western ledge of the basin (Colorado state). A small quantity of oil is obtained from the shaly clay Pirr formation of the Okho area in the Roton basin (New Mexico and Colorado states). As a result of exploration and prospecting over recent years in the Denver basin, the Niobrar formation of Upper Cretaceous age was recognised as a promising oil- and gas-bearing series. Gas flows amounting to 61.6 and 21.1 million m^3/day were obtained at the boring ranges of 405–412.4 and 471–475 metres respectively [46].

LOS ANGELES BASIN

The Los Angeles basin is one of a group of closely contiguous Californian basins, including the basins of Santa Maria and San Joaquin described above.

The comparatively small valleys of this basin are situated amidst the uplift zones of a young folded system of Californian coastal ridges, bordered by the recent coastline of the Pacific Ocean towards which they open out. The system of coastal ridges trends NW-SE, but in the south zigzags considerably. The wide stretch of coastal ridges is known as the Transverse Ridges.

The Los Angeles basin is bound on the south-west and north by a system of uplifts of the Transverse Ridges. A major fault cuts across the entire western part of the basin from north-west to south-east from which a series of feather fractures branch off at an acute angle with predominantly a normal gravity component of movement. Anticlines and brachyanticlines have developed indirectly near these fractures and lie en echelon to their strike throughout the length of the major fault. The majority of these uplifts are productive.

Exploitation of oil was started in the Los Angeles basin in the Wilmington formation in 1932. Exploitation of the fractured rocks of the basement of this formation was begun in 1945 and of the Middle Miocene (Topang formation) and Upper Miocene brown siliceous, micaceous, shaly clays (Monterey formation) in March, 1968. The Wilmington formation is confined to a NW-striking anticlinal fold split into five blocks by meridional faults. The reservoir is a layered anticline broken up into blocks and is of the tectonically screened type [49].

A minor oil-pool in the shaly siliceous clays of the Middle Miocene (Topang formation) is also exploited in the Inglewood oil-field opened in 1924. The reservoir formation occurs in an anticlinal fold complicated by fractures. In type, the reservoir is analogous to the Wilmington formation.

Oil is obtained from the fractured brown, siliceous, micaceous clays in the Long Beach formation in the central part of the Los Angeles basin, where the productive layers belong to the Lower Miocene (Repetto formation). The formation is confined to an anticlinal fold complicated by faulting. The reservoir is stratigraphically screened. The productive layers occur under the surface of unconformity.

BIG HORN BASIN

The Big Horn Basin lies in an intermontane valley of an epiplatform orogen in Wyoming and Montana states. The two formations under exploitation are Elk-Bench and Garland, consisting of siliceous, shaly, bituminous clays as the reservoir rocks.

The Elk-Bench formation, situated in the northern bench of the basin, occurs as an asymmetric anticline dissected by normal and reverse (thrust) faults. The dimensions of the structure are 12.5 km $\times$ 4 km^2 and amplitude of the fold 1500 m. The thickness of the reservoir rocks of the Mohr formation (Lower Cretaceous) is 15.2 m, with average porosity 13%. The reservoir is a layered, anticlinal, tectonically screened type. Oil migration is well known in the fractured siliceous clays of Amsden (Upper Carboniferous) in this formation.

The Garland gas-oil formation is also situated in the upper bench of the basin. The reservoir is confined to an asymmetric anticlinal fold 13 km × 3.2 km^2, with an amplitude of 780 m. The shaly siliceous clays of the Amsden suite (Upper Carboniferous) form the productive layers. In type, the reservoir is similar to the Elk-Bench formation [38].

SAN JUAN BASIN

This basin represents an asymmetric intermontane depression of an epiplatform orogen extending from north to south, that has been filled with sedimentary deposits ranging in age from the Palaeozoic to the Eocene. The basin is 160 km × 200 km and extends over the New Mexico and Colorado states. Oil is exploited from the fractured shaly siliceous clays of the Upper Cretaceous in the Krome oil-field (Upper Mankos formation), in the Red Mountain and Stony Batt oil-fields (Luis formation), and gas is obtained from the Blanko-Mesaverd (Luis formation). The reservoir is lithologically and stratigraphically screened [47].

In recent years, in the region of the overthrust belt of the northern part of the Rocky Mountains (Montana state), oil flows with a discharge of 19.6 t/day have been obtained from the fractured siliceous clay series of Green Horn (Upper Cretaceous) in the Two Medicine area around Gleisher, reaching depths of 2107–2178 m. This is the first occurrence of an oil-pool in the northern part of the overthrust belt in the USA. Oil was struck in 1961 and initially the output was insignificant. Later, production increased tremendously while working on the gas-saturated reservoir of the Upper Carboniferous layer at a depth of 2.512 m. Limestones with intercalations of dolomites forming massive traps of the domal type constitute the condensate gas reservoirs of the formation.

2.2 Gabon Subbasin (Gabon)

The Gabon subbasin forms the end of a huge platformal basin opening into the Atlantic Ocean. The oil-fields in this subbasin are situated on the western border of the delta region of the Ogov River which flows into the Atlantic Ocean a little south of the equator. Amidst the swamps and channels that occupy a large part of the delta, parts of a sandy plain have emerged. One such part is Manji Island, about 50 km long with an area of 500 km^2, extending from Mysa Lopez to the mouth of the Animba River. All the known occurrences of oil and gas in the Gabon subbasin trend north-south in the island and are confined to the region of salt deposits (Fig. 8).

Intensive tectonic movement along the boundary of the Lower and Upper Cretaceous enabled the formation of a series of vast thickness (2000 m) complicated by conglomerates, arkoses and shaly clays (Upper Kokobich formation). As a result, a marine transgression led to the accumulation of thick saline formations intercalated with other marine deposits in the Gabon subbasin. The age

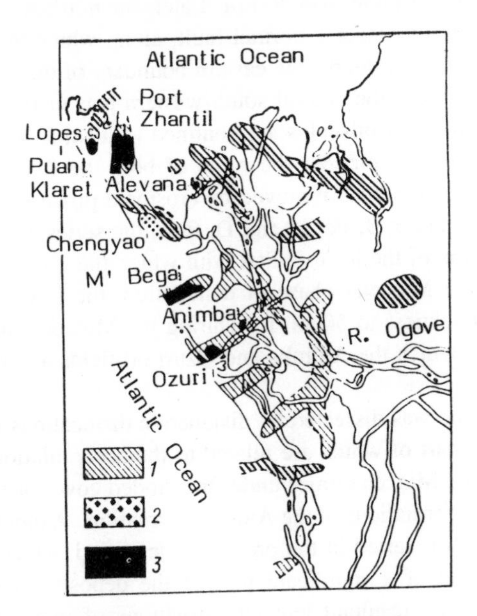

Fig. 8. Schematic map showing distribution of salt diapirs and associated oil- and gas-pools in the Ogov River delta, Gabon subbasin (after J. Pegan and D. Reir).

1—salt diapirs, 2—gas-fields, 3—oil-fields.

of this complex ranges from Upper Aptian to Miocene.

In the middle of the Lower Eocene a sharp change in the character of sediment deposition took place. The deposits of the Ozuri formation formed first with clayey rocks containing increased silica and nodules of radiolarian oozes. The vertical movement, begun in the Cenomanian, facilitated the formation of sandy banks in the underwater gently sloping shore. On these banks originated the peculiar clarified clays which, as shown by J. Pegan and D. Reir, formed an ideal site for siliceous organisms to thrive in large numbers. Thus, the clay complex formed in differently enriched opal and siliceous shaly clays with nodules of clayey dolomites. The non-plastic rocks of this complex explain their disintegration under the action of tectonic stresses and the formation of productive pockets in the Ozuri formation, comprising the Animba, M'bega, Point Clarett, Ozuri and N'chenru oil-fields.

M'bega oil-field is confined to a salt diapir extending from north-west to south-east and continuing under the ocean. The fold developed quickly in the beginning of the Eocene but later its various parts developed unevenly. Maximum uplift took place in the sea. At the beginning of the Miocene, advancing from west to east, the Eocene sediments were washed out into the arched part of the uplift at a depth of more than 300 m. Later, the north-western part of the fold was dissected by a transverse normal fault, along which the north-western part was lowered. This fault marks the eastern boundary of the M'bega oil-field. An analagous fault divides the coastal south-western part of the fold.

The Animba and Ozuri oil-fields are confined to local folds made complex by a gently sloping salt diapir. At the roofs of both the folds a considerable part of the Ozuri series has been removed by erosional processes. According to the information provided by J. Pegan and D. Reir, the series are well preserved only in the lower parts of the local uplifts with which the oil accumulations are associated. In the pre-Miocene phase of denudation, the Eocene Animba formation reduced in thickness to 50–60 m. During the Miocene the development of local uplifts, with which the Animba and Ozuri oil-fields are associated, took place differently.

The Animba dome was dissected by disjunctive dislocations into a series of downthrown blocks, part of which are related to the accumulation of heavy oil, and formed during pre-Miocene times under the eroded cover of small thickness (50–60 m) the clayey formations of the Animba formation. A quick development of the local Ozuri uplift started in the pre-Miocene period, which corresponded to the massive erosion of a significant part of the deposits of the Ozuri and Ikando formations. The resultant lensoid formations of intercalated siliceous shaly clays buried under the weakly permeable Lower Miocene deposits thus became the reservoirs in which oil collected during the post-Miocene period. During the Middle Miocene the Ozuri dome was split by two large cross-cutting faults. The subsidence of the western part (to which one of the productive layers is related) originated at a depth of about 350 m along the line of the transverse fault.

Unlike the remaining oil-fields of Manji Island, the oil-bearing N'Chenru formation is situated not over the roof of the salt dome anticline, but over the far distant western flank. The reservoirs are represented by two horizons of siliceous shaly clays truncated by an unconformably resting Miocene deposit. Gas from this field is utilised for recompression of the M'bega oil formation.

Studies on the reservoir properties of the productive deposits of the Ozuri formation are not complete. In some oil-wells of the M'bega field the fracture porosity varies from 4 to 6%. The fracture porosity values of other oil-fields were determined through analysis of electrical loggings. The results obtained (in %) are as follows: Animba–2.6, Puant-Clarett–2.3 to 7.3 and Ozuri–3.8 to 6.7.

The productivity of the rocks of the Ozuri formation is influenced by fracturing, which, in turn, depends on silicification and OM content, i.e., the textural inhomogeneity. The uneven distribution of these components in the rocks decides their varied reservoir properties both in plan and in section. Related to the predominance of vertical faults, a fracture system showing a tendency towards vertical orientation is observed. According to the data from laboratory measurements taken by J. Pegan and D. Reir, the permeability of the fractured parts goes up to 1 μm^2.

The thickness of the productive zones varies from a few metres up to 60 m and usually increases from the flanks to the roofs of the domes where the most highly productive wells are situated. The oils are heavy (density–0.93 to 0.98) and usually viscous (50–500 mPa).

The peculiarity of the oil-fields in this zone is that the salt massive, forming a diapir with which the formation of oil is associated, does not pierce the overlying complex, in particular the rocks of the Ozuri formation. Tectonically screened oil pools primarily take advantage of this situation for spreading. The oil formations of the Gabon subbasin are usually small. In 1974, the output of oil in the different fields was as follows: M'bega–0.139 million tonnes, N'Chenru–0.176 million tonnes [38].

2.3 Sicilian Basin (Italy)

A petroliferous formation is known in the Sicilian basin where, again, clayey rocks serve as the reservoirs. The Jena oil formation, situated in the south-eastern part of Sicily on the shores of Tunis Bay, was first struck in 1956 and forms the second largest field after the Raguz oil-field. The reservoir of oil–brownish, finely laminated shaly clays (Lower Lias-Middle Trias)–is situated at a depth of 3000–3328 m. The thickness of the oil-bearing part of the dark shaly clays varies from 50 to 150 m; it decreases eastwards, leading to the thinning of clays, which are mixed with limestone-dolomitic rocks. The Mesozoic deposits–from Lias to the Upper Cretaceous –overlie the shaly clays. As shown by core samples of the borewells, a layer of finely elutriated clays, 10 m thick, indirectly rests over the oil-bearing rocks and screens or traps the oil in the reservoir.

According to the information given by T. Rokko, the oil-pool belongs to a gently dipping asymmetric brachyanticlinal fold situated on the flange of the Iblei plateau regionally subsiding north-westwards. The fold axes plunge south-eastwards. The south-western flank of the fold is complicated by a normal fault and a second-order anticlinal closure which, like the principal dome, also contains oil. The fold is 5 km $\times$ 2.5 km, height 379 m.

The most uplifted part of the Iblei plateau appeared first in the Mesozoic part of the Jena formation. Later, starting from the Eocene, to which the Kaltanizett closure is related, it slowly became attenuated. However, subsequent subsidence of the territory exerted no conspicuous influence on the formation of the gently dipping folds, which developed during the pre-Middle Jurassic period. Activation of tectonic movements corresponded with the formation of gravity faults, which facilitated an excellent structural closure of the prospective oil-field, fracturing of the rocks and subsequent migration of oil into them.

The thickness of the Mesozoic deposits diminishes north-westwards. In the region of the Jena formation a significant thinning of the layers lying above and below the Middle Jurassic unconformity is observed. Unconformity as well as local volcanic activity are observed only in the Lower Middle Jurassic.

The following conclusions have been drawn, based on an analysis of studies on the structural properties of oil-bearing formations in countries abroad, where clayey reservoirs also constitute productive zones:

— All oil-producing formations possess clayey reservoirs and are confined to zones of attenuation of monoclines or anticlines broken up into blocks.
— The characteristic type of oil-formations (reservoirs) are layered, folded, tectonic, stratigraphically and lithologically screened.
— In the majority of oil-wells, to ensure adequate inflows or their significant increase, artificial measures have been adopted, in particular blasting, using 80% gelatinised nitroglycerine.
— Testing of wells is conducted in uncemented boreholes at all intervals because the productive zones occur at different levels.
— Commercially productive clayey reservoirs are characterised by a poly-component aspect, with the principal rock-forming minerals being clay minerals (predominantly hydromica), free silica and OM.
— The most prospective wells are situated at the intersections of horizontal and vertical faults, which has resulted in the formation of differently oriented fractures in these rocks.
— Oil and gas-pools in clayey reservoirs constitute one of the stages of the multilayered formations, often one from the lower divisions.

3

Clayey Reservoirs of the Oil- and Gas-Bearing Regions of the USSR

3.1 Domanikian Horizon of the Volga-Ural Region

Most of the Domanikian deposits of the Upper Devonian of the eastern part of the eastern European platform have already been drilled for oil. They have been studied for many years by petroleum technologists and geochemists as potential sources of oil. In recent years, oil was obtained from these deposits in the Tatar, Bashkiriya, Kuibyshev and Perm districts, which has revived interest in them. The various aspects of study of the Domanikian rocks have been extensively discussed in many publications.

The Domanikian horizon is complex, being composed of dark brown or occasionally black, fine-grained, thin, laminated clayey carbonate rocks, enriched in OM and silica, with abundant and highly varied faunal contents wherein the planktonic pteropods, conodonts, goniatites and radiolarians are predominant. Benthic fauna are represented by gastropods and some brachiopods. The Domanikian horizon is distinguished from the overlying and underlying rock layers by the absence of algae, sponges, crinoids, corals, bryozoans, stromatolites and the almost complete absence of foraminifers.

Fluctuations in the proportions of the major components, lithological attributes and facies setting are characteristic of the bituminous, clayey, siliceous carbonate deposits of the Domanikian horizon. S.V. Maksimova distinguished three types of Domanikian deposits according to their composition: the Timanian, southern Uralian (eastern Bashkiriya) and western Bashkiriyan.

Only an insignificant amount of clayey material is contained in the Timanian deposit. Lithologically, it is represented by the following rock types: silicolimestones, siliceous limestones and silicites (Table 3).

The quantity of OM in Domanikian rocks of this type correlates with the content of clay in the rocks. Pure limestones contain the least OM and the amount of C_{org} in them rarely exceeds 3%. N.M, Starkhov, K.F. Rodinova and E.S. Zalmanzon, after a study of core samples from the wells aligned along the profile Teplovka-Kargaly, similarly pointed out that the maximum content of C_{org} is found in rocks with an increased quantity of finely dispersed material.

Table 3: Average contents of components in the Domanikian deposits of the Timanian type

Lithotype	Contents, %		
	Clayey matter	Free silica	Calcite
Silico-limestones	13.66	34.10	26.60
Siliceous limestones	2.50	15.04	78.05
Silicites	4.56	42.68	42.31

According to their data, the average content of C_{org} in rocks of this type is 6.09% versus 0.29% in pure limestones.

The high content of OM in rock samples from the Domanikian horizon has been established by L.A. Gulyaeva and T.N. Gamayunova, who found a high degree of bituminisation of OM[1], with values ranging from 6.27 to 98.8% (average 21.9%), in which the higher the degree of bituminisation, the greater the ratio of clay fraction and OM and the less the C_{org} in the rock.

A similar pattern was also obtained by N.M. Strakhov and his colleagues. However, N.M. Strakhov attributed the differences in degree of bituminisation of OM to the facies conditions of accumulation of OM, particularly the degree of carbonation of the rocks.

The conclusions of these authors have been found correct in two cases. Clayey and finely dispersed carbonate minerals, as shown by our experiments conducted in 1964, 1965 and 1982, have the capacity to adsorb, in other words, to process OM existing in the form of true or colloidal solutions. Carbonate minerals in fine fractions behave like clay fractions of the sediment. In addition to the related acid products formed during transformation of OM, carbonate minerals also facilitate a change in accordance with their exchange capacity value subsequent to the processing of OM.

The Domanikian deposits in the southern Urals (in eastern Bashkiriya) contain far more clayey matter than the Timanian–from 4.13 to 26.38%. However, these values in absolute scale are not that high since in four samples tested by C.V. Maksimova, clayey matter of more than 20% was observed. These Domanikian deposits, compared to the Timanian are younger and characterised by the presence of dolomite in some part of the silico-limestones. The amount of C_{org} is not high in rocks with increased clay content; the value ranges from 4.9 to 13.35% and in limestones with no clayey material or with an insignificant quantity, the C_{org} varies from 1.39-3.62%. For the Domanikian deposits of the southern Urals, these values are sufficiently characteristic but in certain other locations in this region, as shown by N.M. Strakhov, huge lenses of rocks ex-

[1] The degree of bituminisation is expressed in percentages of bitumen carbon from the general content of C_{org}.

tremely enriched in OM are encountered. For example, the lenses of combustible shales in Tashkyskan River have been found to contain 24 to 58% C_{org}.

The Domanikian deposits of the West-Bashkirian type are more clayey and less siliceous than the Timanian and also the Southern Urals. Clayey material constitutes 20 to 30% of the rock. For example, the three specimens studied by C.V. Maksimova contained less than 10% free silica. The clay content increases and the silica content decreases from east to west in an orderly fashion. This relationship in the mineral constituents of the Domanikian rocks reflects the differences in the material characteristics obtained in the sediments, because identical groups of rocks were compared.

A distinct property of the Domanikian rocks (the river flows over the rocks of the Domanikian horizon and consequently deposits of the Domanikian type have not been examined in terms of their genesis or composition) is the high content of finely dispersed (microgranular) calcite of inorganic origin. Further, in the limestones highly enriched in faunal remains the content of non-shelly microgranular carbonate reaches 30-45% of the rock. In the microgranular limestones with radiolarians, finely dispersed calcite makes up 60-65% of the rocks. According to the estimates of S.V. Maksimova, about one-third of the entire mass of sediments of the Domanikian period is accounted for by the quantity of fine-grained carbonate.

Another important aspect of the Domanikian rocks is the presence of free silica and OM in rock-forming proportions. The maximum C_{org} content is typically found in rocks with increased clayey material. An inverse relationship exists between free silica and clayey material.

Silica is present in the rocks of the Domanikian horizon both in concentrations (lenses and intercalations of silicites) and scattered. Scattered silica occurs in a variety of forms. In addition, there are newly formed small crystals of quartz and nodules of variously recrystallised chalcedony (Fig. 9) and distinctive coats on clayey, organic and carbonate (Fig. 10) bodies. Silica occurs as in-fillings in the inner cavities of shells in which it replaces calcite. Further, silicites form small nodules 2 to 7 cm thick and 80-100 cm to a few metres long. They exert no influence over the adjacent overlying and underlying layered deposits.

There is no unanimity of opinion amongst investigators regarding the origin of silica in the rocks of the Domanikian horizon. One group of scientists (L.A. Gulyaeva, N.M. Strakhov and others) considers the source to be silicic acid colloidal solutions derived from the continent, which underwent intensive chemical weathering in the rocks. Data on the forms of iron entering the water are considered indicators of silica saturation. However, forms of iron cannot serve as indicators of significant ingress of silica into the Domanik basin from sources of disintegration because, as is well known, the waters of the seas and the oceans were unsaturated in silicic acid throughout the entire Phanerozoic history of the Earth, when biogenic extraction of silica from the solutions oc-

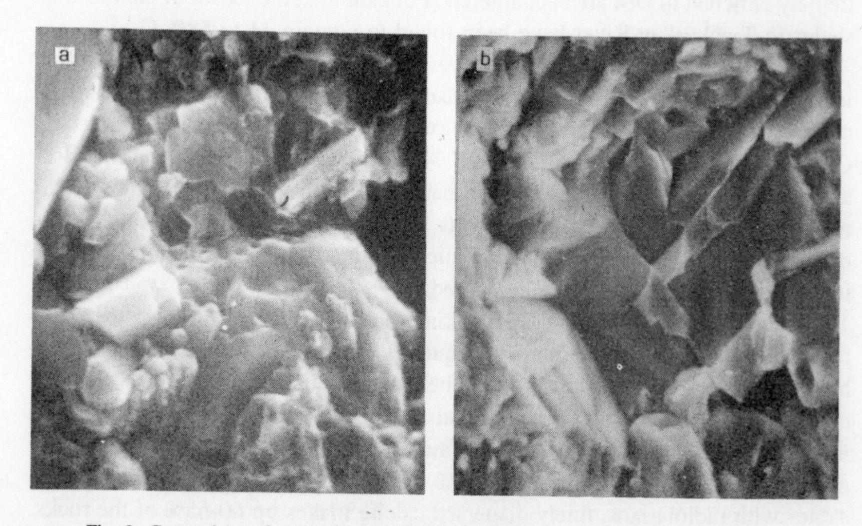

Fig. 9. Concretions of variously recrystallised chalcedony in the rocks of Buraev area (magnification 1000):

a—well 55; depth 1856.6–1860.9 m;
b—well 48; depth 1862.2–1868.7 m.

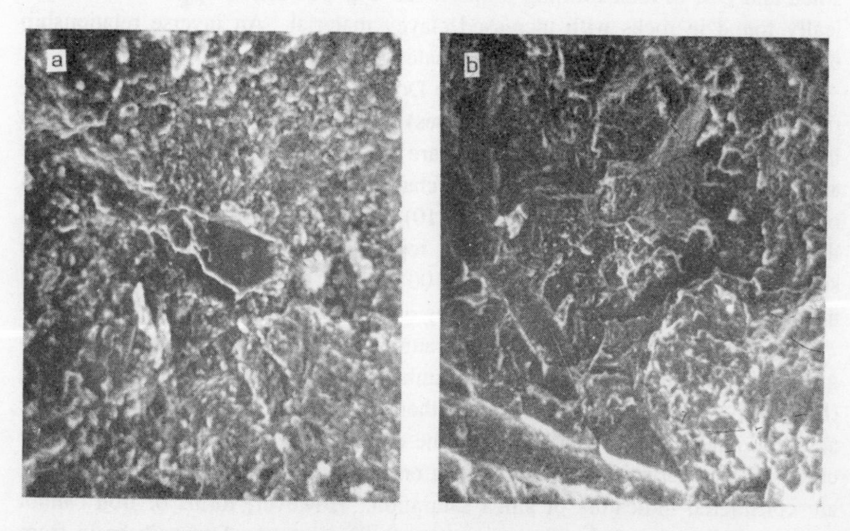

Fig. 10. Siliceous coats in clayey and carbonate bodies, Buraev area, well 48, depth 1862.2–1868.7 m; magn. 1000.

a, b—parallel (a) and perpendicular (b) to the layering of the rock chip.

curred. Furthermore, as shown by G.I. Bushinskii, precipitation of silica has not been established in recent water reservoirs, nor in river estuaries. Siliceous skeletal remains of organisms accumulating at the bottom of the sea show inception of dissolution, during which the silica released becomes diffused in solution and might either enter the biogenic cycle, or during supersaturation of the solutions in the acid medium, carbonic acid is formed (which is released during the processing of OM of the second type), precipitating silica in the sediments.

Another group of investigators (S.V. Tikhomirov, V.A. Zav'yalov, S.V. Maksimova and others) invokes centres of vast underwater vulcanicity as the source of silica.

The bloom of fauna with siliceous skeletal parts, the significant silicification of the Domanikian rocks, and also the presence of a vulcanogenic siliceous formation of the same age in the Uralian geosyncline, support with no shadow of a doubt the presence of an endogenous undercurrent of silicic acid in the Domanik basin.

As regards the stage of silicification of the rocks of the Domanikian horizon, N.M. Strakhov, S.V. Tikhomirov and S.V. Maksimova suggest the presence of two major stages in this process. The first stage, dispersed silicification, is associated with the initial stages of diagenesis when the sediments were still sufficiently flooded. At this time, a hard skeleton with associated deformational characteristics of the heterogeneous parts formed in the clayey part of the Domanikian rocks and entered the composition of the deposits. The second stage, concentrated silicification, is related to the later stages of diagenesis and passed through a considerably thick pile of sediments. At this stage, the various kinds of concretions of silica took the form of layers or large lenses.

Besides the above-mentioned two major stages of silicification, belonging to different stages of diagenesis, there also existed a third or katagenetic stage of silicification of the Domanikian rocks. In these rocks, inside the sheeted lenses of silicites and coloured OM, shells showing replacement by silica and uncoloured OM were encountered. To this stage belongs the formation of concretionary calcite and kaolinite and uncoloured OM. The katagenetic stage of silicification is quantitatively insignificant compared to the first two stages, but is quite important in that it accounts for the enrichment of pore waters in silica.

In 1982, G.I. Surkova distinguished three generations of post-sedimentation silica in the rocks of the Domanikian horizon. The earliest separations of silica are in the form of globules and fine granular cumulates supporting its precipitation as colloidal formations from solutions in the flooded sediment. Silica of the second generation quantitatively changed from weak silicification of organic limestones to continuous transformation of carbonate rocks into silico-limestone and silicite. The organogenic, more porous and permeable rocks, sufficiently consolidated over a period of time for the release of silica of the second generation, were subjected to the most extensive silicification. Silica of the third generation developed both in the cavities and through dolomitisation and considerable recrystallisation of the limestones. To the silica of the third generation is also related chalcedonic filling of the fissures in the siliceous combustible shales, silico-limestones and silicites.

All the three types of OM described in Chapter 1.3 are present in the rocks of the Domanikian horizon. Certain differences are noticed, however, in the distribution of OM in the rocks. Organic matter of the first type is encountered in all Domanikian rocks with heterogeneous facies. It is distributed relatively uniformly, only occasionally forming layers, with fully transformed globular pyrite. Occasionally, along with OM of the first type, weakly recrystallised clusters or well-cut cubes of pyrite are found. In general, OM of the first type is the major source of pyrite in the Domanikian rocks.

Organic matter of the second type is encountered in the form of vegetal remains and collomorphic clusters. Depending on the quantity of clayey minerals present in the rock, it is either localised in the clayey layers (Fig. 11) or fills the various types of cavities in the carbonate part of the rock. In the process of diagenetic transformation of the humus part of OM of the second type, the breaking of oxygen links took place, giving rise to free carbon dioxide and lowering of the pH of the pore waters. As a result, the carbonate minerals lying in direct proximity to OM dissolved. The humic and sapropelic components of OM of the second type simultaneously released hydrolysed components, which were adsorbed by clayey and carbonate minerals, particularly of a fine-grained nature, and also filled the free space during the solution of carbonate minerals. Sometimes, at some distance from the organic remains, newly formed anatase (single crystals), pyrite and marcasite are observed. The last crystallises only under the condition that large fissures or caverns exist in close proximity to an OM accumulation of the second type.

Organic compounds of any kind belong to OM of the third type, if they exist in the form of true or colloidal solutions. This OM is adsorbed by clayey and carbonate minerals, preventing any change from their natural state in composition or degree of crystallinity. Thus in the Domanikian rocks, parts of fine-grained limestones that adsorbed OM have remained unchanged while those without OM show significant recrystallisation.

The mineral composition of the clayey part of the Domanikian rocks is almost identical for all the distinguished types. The main clay mineral is dioctahedral hydromica. Mixed-layered minerals of the hydromica type–montmorillonite, belonging normally to thin clayey layers inside the siliceous-carbonate matter, are present in the form of impurities. Neither hydromicas nor the mixed-layered minerals give sharp reflections in diffractograms in spite of routine processing (Fig. 12). In some samples the reflections of hydromica and mixed-layered minerals are hardly registered. The reason lies in the adsorption of OM and silica by the clay minerals. In rare

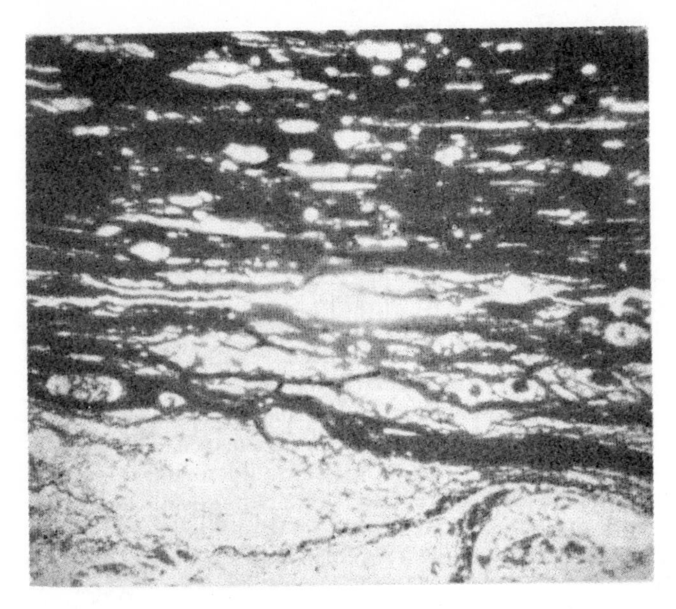

Fig. 11. Laminated texture of the rocks of the Domanikian horizon with OM of the second type. Tulva area, well 76, depth 2086.5-2089.5 m; magn. 40, Nichols II.

samples, enriched in OM of the second type, the transformation, as mentioned earlier, is accompanied by the release of carbonic acid and the formation of kaolinite from the parts of hydromica, which is unstable in an acid medium.

An interesting characteristic of the clayey component of the Domanikian rocks was noticed by P.A. Gulyaeva and her colleagues, namely, the absence in them of the minor elements, while the increased content of vanadium, nickel and copper in the rocks, in their opinion favoured active accumulation of these elements in OM from the waters of the basin, where they should have reached in the form of true solutions. All this clearly indicates the endogenous source of their entry into the system.

The deposits of the Domanikian horizon are sharply differentiated in the general cross-section of the carbonate part of the Devonian, both through electrical characteristics and the complex faunal content. They are characterised by significantly high apparent resistivity (AR) (average of 500–1000 ohm·m) which serves as the reference point in carrying out correlations of the sections above and below. In the geoelectrical curve of the Domanikian logs, in the majority of cases, the peaks correspond to a single maximum of the AR high and sharp falls in the curve of potential reflect the characteristic polarisation (CP). Still, the configuration of this peak and the magnitude of AR, as shown by L.I. Sokolova, are not constant. A single peak is divided into two, three or more

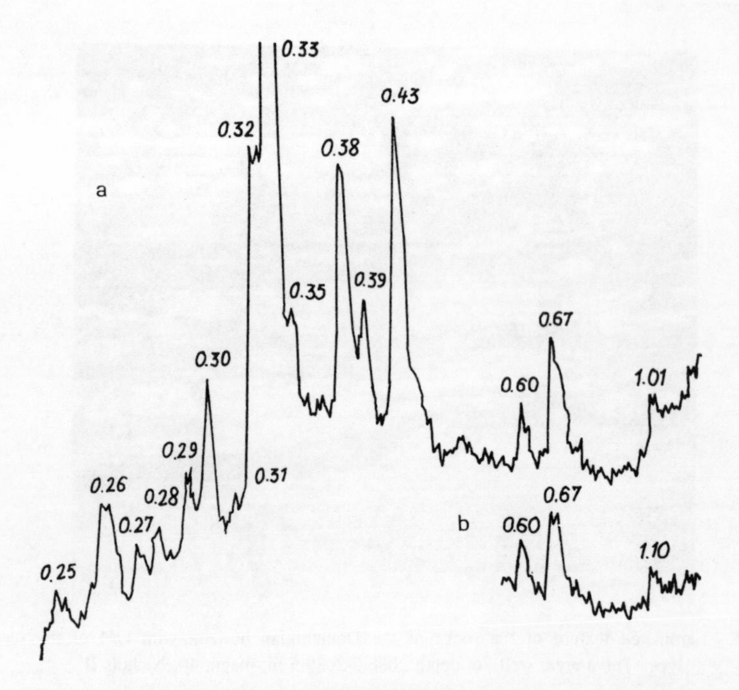

Fig. 12. Diffractograms of rocks of the Domanikian horizon (a—original sample, b—sample soaked in glycerine). Buraev area, well 55, depth 1856.6-1860.9 m. Numbers—intensity of mineral reflection, nm.

maxima as the observation point is shifted from east to west and from south to north. This reflects the change in the material content of the Domanikian horizon in these directions.

The rocks of the Domanikian horizon form a belt amidst continuous marine deposits conspicuously differing from the sediments of the adjacent zone both lithologically and in their faunal content characteristics. Analysing the development of the Russian platform in the Devonian, S.V. Tikhomirov showed that at that time the eastern part of the platform differed from the central part. The meridional Eastern Russian depression appeared and stood out prominently in the first half of the Frasnian period. This structure formed during the subsidence of the eastern part along a system of faults worked out by N.S. Shatskii as the 'Main Fault'. The Eastern Russian depression, with its distinctly flat terrain structure of the second order, presented a typical zone of subsidence where, at the time of transition, according to the data given by S.V. Maksimova, the geomorphic relief was prominent because of the presence of typical scarps and steep slopes with a sheer drop of 50 to 100 m.

As shown by M.F. Mirchink and others [33], the time of accumulation of sediments was an important geological boundary dividing the various stages of development of the Volga-Ural'sk region according to tectono-sedimentation conditions and structural morphological characteristics of the sediments. The thickness of the Domanikian rocks varies from 5 to 80 m, forming certain regional maxima and minima of varying configuration. The minima correspond, in most cases, to the position of recent swells, the depressions of which divide the linear anomalies of increased thickness.

The structural differentiation of the Domanik basin together with the remnant features of the tectonic setting of earlier periods, is characterised by distinctly newly formed features which occupied a large (eastern) part of the Urals-Povolzh region and led to the appearance of structural facies zoning in the Domanik basin, superimposed to a large measure, over the palaeostructure of the more ancient basin. To the newly formed elements of this part of the basin belong the Shkapov-Belebeev palaeo-anticline, appearing in the boundaries of the eastern part of the Upper Proterozoic of the Sergiev-Abdulin graben depression and the Krasnokam-Polaznen horst (bank), evolving together as the most subsident axial zone of the Upper Kam depression of the Eifel-Pashii basin. The wide zone of the Mrakov-Krasnoufim anticline, occupying the eastern margin of the platform, evolved up to the end of the Sargaevian period. It is dismembered into a series of smaller palaeo-anticlines (for example, Kungur-Kynov and Bashkiriya) which are subdivided into the sublatitudinal Kokui-Serebryan and Aktanysh-Cheshmin depressions.

Thus the Domanikian period was one of active tectonic development in the Urals-Povolzh region, and not of relative quiescence, as shown by many earlier workers.

The most frequent expressions of the faults of the basement in the sedimentary cover are the flexural folds of numerous linear structural arches and other uplifts. The most active systems of tectonic fracturing of the stressed zones of the sedimentary cover are related to these zones and appear at the contacts of the blocks characterised by multidirectional movements [33].

Considering the small thickness of the rocks of the Domanikian horizon and their relatively uniform lithofacies character, many workers believe that the inadequate amplitude of tectonic uplift of the bottom of the Domanik basin through sediment accumulation is betrayed by the incomplete compensation of such uplifts by the sediments [3, 33].

The palaeostructural differentiation of the basin of the Domanikian period favoured conspicuous differences in the facies setting of the sedimentation in it.

M.F. Mirchink, O.M. Mkrtchan and A.A. Trokhova, in reconstructing the palaeostructural development of the Domanik basin, established direct dependence of the faults on the thickness of the sediments of the Domanikian period from the nature and amplitude of the tectonic movements. These authors analysed the regional changes in the thickness and lithofacies of the Domanik basin of the large palaeostructures (anticlines, gently sloping uplifted blocks, depressions and troughs) formed during the synsedimentary tectonic differentiation of the basin. The diverse relationships between thickness, lithofacies characteristics and intensity of tectonic movements highlight the varying degree of compensation by the sediments subsiding in consequence of the fault zone activity. It is important to note the unevenness of the relief of the bottom of the Domanik basin, which is related to the formation of interlayers of shallow calcareous dolomites and lenses of pure dolomites in the gently tilted uplifted blocks.

In the Domanikian horizon organic and biochemogenic limestones predominate. Amongst the organic limestones the most widely distributed are the polydendritic varieties. Biochemogenic limestones, composed primarily of micro- and fine-grained calcite, are normally bituminised and form thin intercalations with clayey limestones, marls and bituminised shaly limestones. All these types of rocks are enriched in clayey matter to a varying degree, which plays a dominant role in creating textural heterogeneity, leading to a favourable situation in the rocks for the formation of fractures.

The textural feature of the rocks of the Domanikian horizon arose through the interaction of the four rock-forming components, viz., carbonate (chemogenic or shelly), free silica, OM and clay minerals. The shell limestone seems to float in the form of layers or lenses or individual shells (Fig. 13), forming lenticular or indistinctly laminated mesotextures. Silica also participated in the formation of lenticular-laminated mesotextures. Indistinctly laminated mesotextures are very often shown by the parts of OM of the second type (see Fig. 11). Microtextures appear due to the oriented distribution of fine-grained calcite and clay minerals around centres of stresses which, in the Domanikian rocks, happen to be silica nodules, large crystals of calcite or dolomite (Fig. 14) or fossilised fauna. Microtextures were formed by parts of OM adsorbed by the clayey and carbonate minerals (Fig. 15). Sorption of OM of the third type was accompanied by hydrophobisation of the clayey and fine-grained carbonate minerals. As mentioned earlier, the contacts between hydrophobised surfaces and those of hydrophobic surfaces with hydrophilic surfaces are far weaker and thus more liable to breaking and disintegration than the contacts between hydrophilic surfaces.

The diverse mineral and organic components participating in the formation of the textural characteristics of the reservoirs of the Domanikian horizon gave rise to a variety of sizes and forms of contact. The special feature of the post-sedimentation history of these rocks, which contain a considerable quantity of chemogenic carbonate matter, is the quick dehydration of the sediments and shorter duration of the diagenetic stage. Therefore, the final formation of the textures and the weakened zones along the borders took place at the last stage of diagenesis, not much beyond the consolidation of the sediments and their conversion to rock. Between diagenesis and katagenesis considerable time elapsed during which the rocks attained a stable condition.

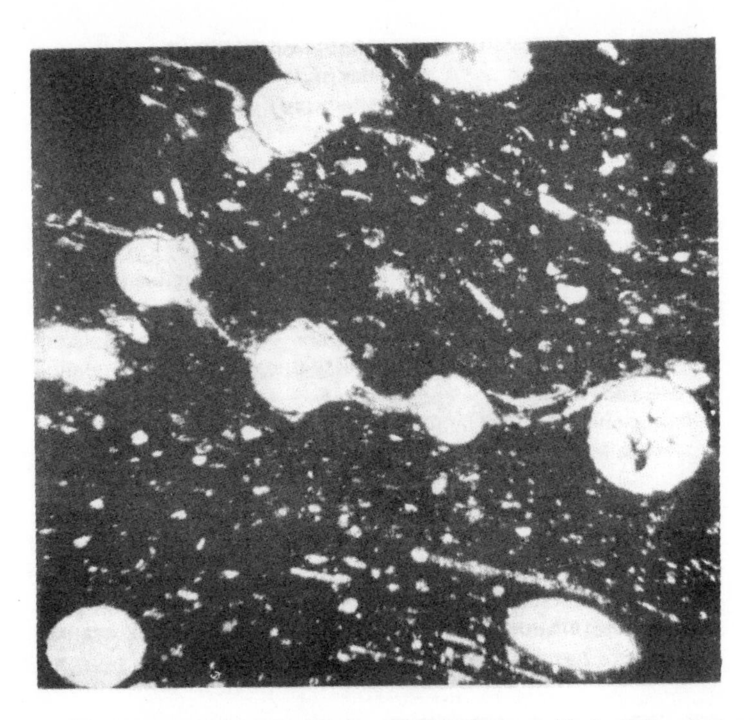

Fig. 13. Mesotextures formed by the distribution of recrystallised shells of conodonts and the binding calcite matrix (Tulva area, well 76, depth 2086.5–2089.5 m); magn. 150, Nichols II.

The change of thermodynamic parameters related to the transformation in the zone of katagenesis corresponded to the separation of parts of rocks along the textural borders. This means that, at that instant, conditions were most favourable for the reservoirs to be filled with oil and gas, i.e., for the formation of oil and gas-pools in the clayey reservoirs [19].

The area of weakened zones at the mesolevel is well discernible in thin sections and is considered indicative of fractures, for which the method adopted by VNIGRI [21] was employed to determine the volume and filtration parameters. The following data were obtained for the Domanikian reservoirs: opening of the fracture, i.e., width of the weakened zones, up to 1 mm; volume of fracture area, from 50 to 400 m^{-1}; fracture (average) porosity, 0.2 to 0.3%; and permeability, 0.010–0.015 μm^3 [12].

It is desirable to note again that the terms 'weakened zones' and 'fractures' are not synonymous. For a fracture to form in rocks during tectonic stresses, the rocks ought to be brittle. Weakened zones appeared in plastic rocks as a result of the contacts between various parts according to their texture. The reservoirs

Fig. 14. Reorientation of clayey and carbonate particles by a series of large crystals of calcite, Buraev area, well 55, depth 1856.6–1860.9 m; magn. 300.

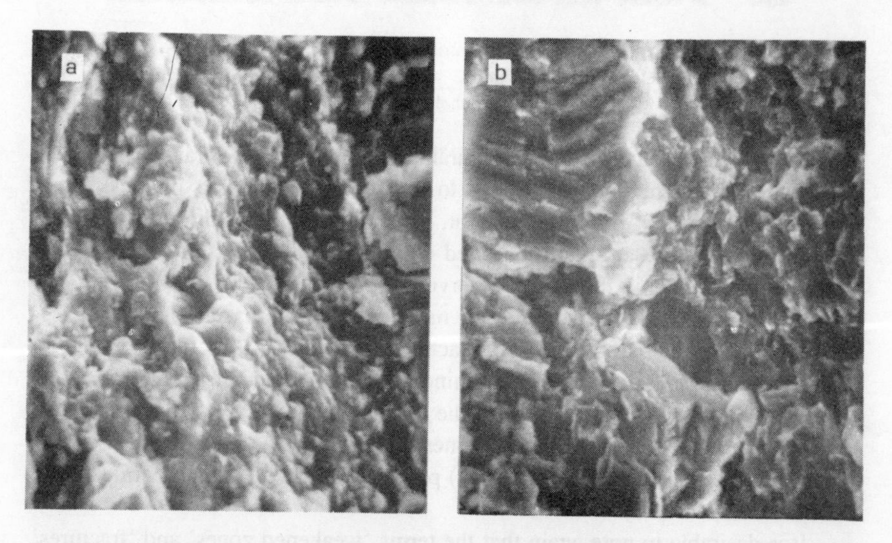

Fig. 15. Microtexture of rocks of the Domanikian horizon of Buraev area, with OM of the third type (magn. 500). a—in siliceous-clayey part of rock (well 48, depth 1862.8–1868.7 m) and b—in siliceous-carbonate part of rock (well 55, depth 1856.6–1860.9 m).

of the Domanikian horizon consist of parts of diverse competency. In parts consisting of chemogenic calcite without an admixture of clay, the appearance of fractures is possible due to the disruption of continuity of the brittle part of the rocks under the action of pressure on certain other acute angle fragments of the rocks. But to date such parts have been so small that they could not be considered without error while attempting a precise prognosis of reservoir properties. Although the separation of weakened zones as well as fracturing led to crushing and brecciation of the rocks, the mechanism of this process cannot be isolated and delinked completely from the loss of plasticity by the rocks. If the clayey reservoirs of the Domanikian horizon were not texturally heterogeneous, they would not have served as reservoirs with conducive volume and filtration parameters.

The separation of texturally different parts of the rocks (Fig. 16) becomes more pronounced in the series of changes induced by loss of water and the sorbate clayey and carbonate (fine and microgranular) components in them. These changes did not take place simultaneously. First, a small film of water was lost through adsorption by the carbonate minerals present in those places where there were no clay minerals. The rocks began to separate in these places, both along the zones of contact of different rock fragments and along the weakened parts of the same rock fragments. The water molecules lost by the clay minerals and subsumed in the composition of the rocks of the Domanikian horizon, led to the weakening of links between carbonate and clay minerals and their separation.

It is necessary to note that the linkage of clay minerals with carbonates does not differ in strength. The presence of a thin film of clayey matter between large carbonates, particularly the shell fragments, as shown by our experiments, increases the tendency of this part of the rock to separate from the adjacent matrix. If the clay minerals form a fine mixture with the chemogenic carbonate material, then this clayey-carbonate mass together possesses a higher plasticity and hence a larger durable link with the shell carbonate fragments than each of its components individually. The zones of contact of the clayey-carbonate mass with organogenic fragments are less conducive to dissociation than the zones of their contact with clayey minerals or fine-grained chemogenic carbonates.

The important characteristic of the formation of the useful capacity of the clayey reservoirs of the Domanikian horizon was the opening of pores during the separation of weakened zones. As revealed by a study of thin sections of these rocks, the boundary zones with various textures cut across the system of pores, the sizes of which depend on the composition and grain size of the contact parts. The dissociation of weakened zones led to an increase in the dimensions of pores to a size equal to the width of the interstice being formed. At the same time, the pores not only increased in size but also became interconnected through the weakened zones by means of large canals that widened. This feature obviously sharply bettered the filtration characteristics of the rocks.

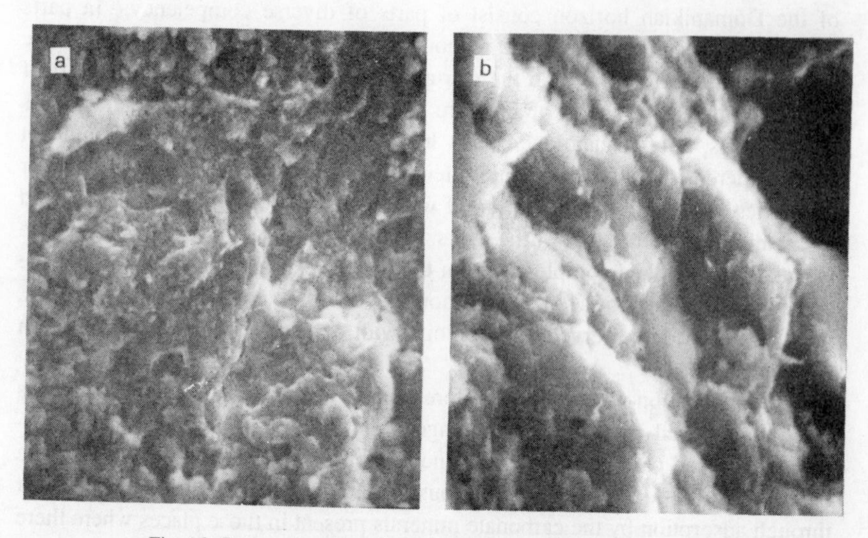

Fig. 16. Separated weakened zones at the microlevel (magn. 1500):

a—Buraev area, well 55, depth 1856.6–1860.9 m;
b—Beketov area, well 32, depth 2339.7–2345.2 m.

Three reservoir layers–D^1_{DM}, D^2_{DM} and D^3_{DM}–are distinguished in the cross-section of the Domanikian horizon. The upper two layers are distributed regionally and the third is situated at the bottom of the horizon locally. We have observed that the values of porosity and permeability of these rocks determined from the cores do in fact provide distinct but lower safe discharges of up to 60 m^3/day. The reason for this lies in the fact that the reservoir properties of the clayey reservoirs depend on textural heterogeneity, which favours the formation of weakened zones (already mentioned above). These zones, dissociated under the pressure of the migrating oil during the formation of pools, release oil during the development of a well and open up during the lowering of pressure due to drilling.

The estimated porosity values of the reservoirs of the Domanikian horizon usually do not exceed 5%. The pore dimensions of the clayey reservoirs vary from 0.5 to 1.7–2.3 μm and the width of the weakened zones works out to 5–27 μm. Based on the general dimensions of the weakened zones and their expansion, it is possible, perhaps, to more than merely comment on the order of the useful capacity of this type of clayey reservoirs, when such structures attain conditions conducive to the rifting of the weakened zones.

The most complete data on the oil-bearing characteristics of the Domanikian horizon have been collected from the Tatar oil-field. Reservoirs with good filtration properties are distributed sporadically and belong mainly to the Al'mete'v anticline. Here they possess the following characteristics: porosity, 0.2 to 3–5%; gas permeability, 0.010–0.011 μm^2; discharge, 20–60 m^3/day [3]. Such a type of reservoir exists in the North Tartar dome (Shadchin and Shii oil-fields). Reservoirs with medium reservoir properties (discharges up to 6–12 m^3/day) belong to the Urat'min, Tlyancha-Tamak and Menzelin-Aktanysh areas.

All known accumulations of oil in the reservoirs of the Domanikian horizon belong to the tectonically screened type and are confined to the zones of faults and flexures which correspond to the linear zones of fracturing [3]. The structure of the oil occurrence proved in the eastern part of the Aktanysh-Cheshmin trough is particularly interesting. It belongs to the highly complex overthrust folds of the Kinzebulatov type [33].

Similarly, the useful or exploitable volume of the reservoirs of the Domanikian horizon is favoured by the textural heterogeneity at meso- and microlevels. This arises undoubtedly from the values of porosity and permeability as measured from core samples, which cannot guarantee the very high discharges normally obtained during the working of oil-wells from reservoirs of this type. The measured dimensions of pores and the weakened zones obtained through studies of photographs of thin sections and photographs from SEM over the electronic computer Quantimet-720 emphasise the need to estimate the capacity and filtration properties of potential clayey reservoirs.

The potential of the rocks of the Domanikian horizon ought to be more fully realised by opening up new oil-fields. There are certain favourable indications for this: first, the textural heterogeneity, with parts of silica forming the hard and rigid framework in parts situated in clayey and fine-grained carbonate minerals; second, the presence of faults by means of which oil resources accumulated in the terrigenous Devonian. As shown by M.F. Mirchink and others [33], the structures of the Domanikian period inherited and developed on the earlier established block structure of the basement and the terrigenous Devonian. This means it is essential to explore the Domanikian horizon for such structures as those formed under the influence of fault tectonics, since it is precisely here in the terrigenous Devonian and Carboniferous, i.e., in the layers immediately above and below, that oil lies.

Summarising what has been said about the reservoir rocks of the Domanikian horizon, the following conclusions may be drawn:

- All the principal minerals and structural-textural transformations in the rocks of the horizon were completed at the end of diagenesis, after which the stage of silicification set in, but not beyond the stage of accumulation of commercial oil.
- Rocks of the horizon are enriched in OM of mostly the sapropelic type and its transformation under the influence of high temperature facilitated the hydrophobisation of the zones of contacts of various textures. This process did not take place at one time; initially it took place during

sedimentation and later, during diagenesis.

— Oil-pools formed in the horizon during the vertical migration of hydrocarbons (HC), which is also confirmed by a comparative palynological study of the rocks and oil.
— Rocks of the horizon are considerably enriched in free silica, the predominant part of which is associated with the migration of solutions of silica along the faults in the Domanik basin and in the sediments at the beginning and middiagenesis; additionally, dispersed silicification in the form of microlenses and layers of silica is also evident.
— Rocks of the horizon are distributed over a vast territory and are not thick.
— One-third of the deposits of the horizon, for example, consists of fine granular carbonate material that accumulated at the stage of sedimentation during changes in the cationic proportions because of the flow of material through pre-existing faults;
— Rocks of the horizon contain a significant quantity of products of activity of organisms, including phosphorous, related to the detritus of fish. The siliceous organisms are mainly represented by radiolarians.

3.2 Bazhenovian Formation (Western Siberia)

The bituminous siliceous rocks of the Bazhenov formation constitute around 1% of the thickness of the cross-section of the sedimentary cover of the western Siberian platform but are developed over a vast territory (more than 1 million km^2) from the lower reaches of the North Sos'va River in the south up to Omsk-Kollashevo in the west. Their thickness in independent structures is controlled by the basement structure and varies from single units to 50 m. In the relatively narrow belt close to the hilly terrain in the west, south and east, the Bazhenovian deposits change over to the non-bituminous variety of the Mar'yanov formation and later to the calcareous glauconitic siltstones and sandstones of the Federovian, Shaimian and Maksimoyarian formations [9].

Rocks of the Bazhenovian formation were first distinguished by F.G. Gurari in 1959 as a bituminous pocket forming part of the Mar'yanovian formation. Initially, these rocks were considered the major oil-producing formations or one of the most consistent regional covers or traps over the oil and gas horizons of the Jurassic complex. After this, a series of formations were discovered in western Siberia (Salym, Pravdin, Verkhne-Salym, Verkhne-Shapshin and others) with rocks of the Bazhenovian formation, which proved to contain oil resources on a commercial scale. Further, the conditions, formation and characteristics of oil accumulation remained the subject of research for many investigators. As a result, the differences in composition and properties of these deposits in various regions of western Siberia and also the differences between the rocks of the

Bazhenovian formation and the overlying and underlying rocks were ascertained. Because of their specific petrographic features, the rocks of the Bazhenovian formation are sharply differentiated in log diagrams through anomalously high apparent resistivity and natural radioactivity, establishing the most consistent reference point in the sedimentary cover of western Siberia. The Bazhenovian horizon also serves as a reflecting seismic horizon [6], lying at depths ranging from 600 to 4000 m.

Results of studies have shown that depending on the tectonic activity and the characteristics of the rocks intersecting the various facies zones of the western Siberian petroliferous basin, the rocks of the Bazhenovian formation could either be cap rocks or reservoirs. The problem of the petroliferous characteristics of these rocks has not met with a unanimous solution.

The bituminous siliceous rocks of the Bazhenovian formation form, in fact, the boundary layer between the Jurassic and the Cretaceous systems. The extreme boundary is seen as a thin intercalation (2–4 m) of sideritised rocks, characterised by lowered electrical conductivity.

The Bazhenovian formation on the whole is impoverished in organic remains. However, individual intercalations of Bazhenovian rocks contain scales and skeletal remains of fish, cephalapod shells, brachiopod Lingula, segregation of ammonites, belemnites and pelecypods. Microplankton is represented by radiolarians and coccolithoforids. At the central part of the basin the Bazhenovian rocks show maximum impoverishment in faunal content and the complete absence of foraminifers.

The age of the Bazhenovian formation (Volgian stage —Berriasian) has been established on the basis of palaeontological remains. Ammonites have proved to be most significant in the biostratigraphic division of the formation. An analysis of their distribution and that of other groups of fauna along the section made by G.S. Yasovich and M.D. Poplav in 1975 served as the basis for their division of the formation into two members of different ages, the Lower Volgian and the Upper Berriasian. This division is supported by the geological data. The upper stage shows more uniformity in thickness (12–14 m), and the lower stage varies from 10 to 40 m, while its maximum thickness is reached close to the periphery of distribution of the formation. Members also differ in their degree of bitumen and carbonate contents. The maximum bitumen, carbonate and silica contents are seen in rocks of the upper member of the formation.

The carbonate content of the upper member of the formation was increased by the massive accumulation of coccolithoforids, because of which T.I. Gurova assigned these rocks to the category of marls. Our investigations in 1981 showed that in the rocks of the upper member carbonate inclusions are encountered in the form of oolitic (nodular) layers, at times as bituminous limestones 2–3 to 5–6 mm thick, individual crystals (from 0.04 to 0.20 mm) and microlenses of calcite randomly distributed in the clayey part of the rock and fragments of carbonate fauna (prismatic layer of pelecypods). In the lower part of the formation at well 118 opened up in the Salym area an accumulation of crystals of fine-grained dolomite is uniformly distributed throughout the entire thin section and fragments of carbonate fauna are partly silicified. In some samples the finely dispersed clayey mass is overcrowded with globules of siderite forming dark belts running parallel to the bedding. Very large globules are 0.010–0.015 mm in size. These siderite concretions are rimmed by globules of pyrite. Individual rounded concretions of calcite (2–3 nodules in a specimen) are common.

Detailed studies were conducted on the ammonites by M.S. Mesezhnikov [27] to establish middle Volgian and Upper Volgian substages and also the Berriasian stage. As shown by G.E. Kozlova, radiolarians serve as the index fossils in pinpointing the biostratigraphic divisions of the Bazhenovian formation.

Macroscopically, the Bazhenovian formation presents a monotonous series of black bituminous siliceous rocks which, at places, are calcareous and considerably pyritised. A minor admixture of terrigenous material, primarily fine silty or coarse pelitic (from 1 to 3%) is common. Only in the border areas of distribution of the formation (Berezov monocline, Kollashev Pri-Ob′) does its content reach 8% and the grain size increase to that of medium silt. The Bazhenovian rocks differ from the overlying and underlying rocks in their increased contents of OM (5–18%, up to 23% in the Salym area), silica (10 to 30%), high natural radioactivity (40–80, sometimes up to 1000 mkR/hr) and specific electrical resistance (25–625 ohm·m), with the values going up to 4500 ohm·m in the Salym area. These rocks also possess increased general porosity (5.8–8%) and low density (2.4–2.23 g/cm^3). These differences are observed everywhere and not merely confined to the regions where oil is found in the Bazhenovian formation. Thus, according to the data given by O.G. Zaripov and I.I. Nesterov [16], outside the zone of oil accumulation the density of the rocks does not exceed 2.4 g/cm^3 and porosity, 5.8%. Then, as in the overlying clays, at their depths of occurrence, the values correspond to 2.6–2.7 g/cm^3 and 4–4.2%.

The fields of distribution of anomalous values of the cited parameters do not correspond to the boundaries of the Bazhenov basin but form sharp asymmetric shears in the western part of the depression (western meridian of Urengoi—Nizhnevartov). Then, a gradual lowering of silica content, natural radioactivity and other parameters, such as sonic, are observed during the transition to the non-bituminous variety in the border parts of the basin.

I.I. Pluman [34] relates the characteristically high natural radioactivity in the rocks of the Bazhenovian formation to the presence of uranium, the content of which is higher in the rocks of this formation than in those above and below. According to the data given by V.V. Khabarov and others [43], uranium in natural radioactivity contributes 80–90%, potassium 5–10% and thorium 5–10%. These workers believe there is a direct relation between intensity of radioactivity and OM content. Separation of the Bazhenovian formation into stages has been carried out and rocks differentiated by means of gamma ray logging (GL) (Fig. 17) and according to the uranium content [43], 10^{-4}%:

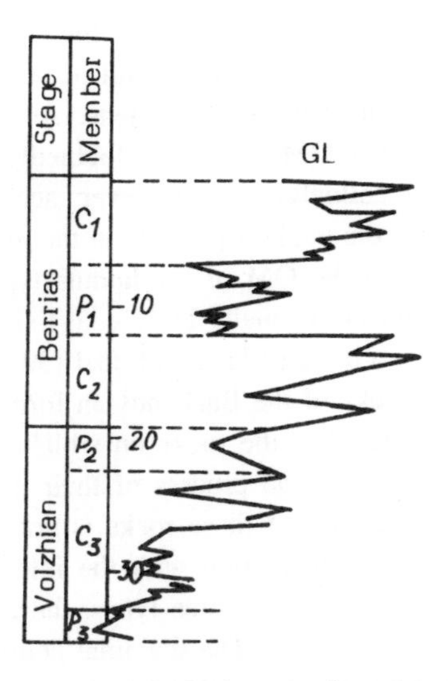

Fig. 17. Penetrated intervals in rocks of the Bazhenovian formation of the Salym oil-field (from V.V. Khabarov, O.M. Nelepchenko and T.V. Pervukhina, 1978).

C_1	56	C_2	59.6	C_3	25.3
P_1	27.7	P_2	20.3	P_3	12.4

The major part of the uranium is present in a barely soluble form and is linked with OM and the phosphatised remains of fish. The mobile form of uranium amounts to 2.10^{-4}% [43]. The maximum enrichment of uranium in the rocks of the central parts of the Bazhenov basin caused I.I. Pluman [34] to posit that the uranium reached the rocks not from a source of predominantly mobilised material, but rather from marine waters under reducing conditions. Most researchers are divided over this conclusion [9, 11, 35]. V.V. Khabarov and others [43] cite the characteristic of maximum extraction of uranium from hydrous solutions and its permanent retention in humic OM and propose that the enrichment of uranium in the rocks took place at the stage of early diagenesis. However, the mechanism of enrichment of uranium in marine waters is not discussed in any of the cited works.

The total content of OM in the rocks of the Bazhenovian formation varies systematically throughout the territory of western Siberia. The maximum quantity of OM (10–18%) is noticed in the structures of the Salym-Surgut region.

The region of increased amounts of OM (5–10%) covers a large part of Nadym and Khanty-Mansii depression and also extends southwards to the Yugan and Omsk depressions. Along the directions approaching the periphery of the basin the quantity of OM drops from 5 to 1%.

The main material for OM was provided by the simplest planktonic organisms (blue-green, green algae and ?), the chemical composition of which was fairly close and quite similar to the sporo-pollens. According to the data from coal petrographic studies [17], OM belongs undoubtedly to the sapropelites and is present in the form of colloalginite. However, according to the results of many petrographic studies, the nucleus [30, 41] in the rocks of the Bazhenovian formation has been found to be OM of the humus type, present in the form of fine vegetal detritus and occasionally large coal-fixing and vitrinite-locking remains and also thin (up to 0.3 cm) layers of coal [30].

Organic matter in the rocks of the Bazhenovian formation is a unique rock-forming component. It penetrates the rocks on which it is present and above all exerts specific limitations on the process of their interactive influence and transformation. Further, the OM of these rocks is not of one kind. It differs according to the character of interaction with the mineral constituents of the rocks and accordingly is divided into three types, the characteristics of which have been described in Chapter 1.3. The maximal content of OM of the third type is associated with the field of distribution of anomalous values of natural radioactivity, apparent electrical conductivity and other parameters.

The universal concentration of the radiolarians in the rocks of the Bazhenovian formation and also the high content of silica were considered by V.P. Kazarinov and T.I. Gurova in designating these rocks as chalcedonic radiolarites. Silica occurs in the rocks of the Bazhenovian formation often as lenticular bodies but in most cases it is found uniformly scattered in the rocks [30].

There is no unanimity of opinion amongst scientists with regard to the principal sources of silica in the rocks and the nature of the processes which led to its accumulation in the sediments. At present, three possible sources of free silicic acid entering the basin of sedimentation are known–erosion of land, volcanic activity and siliceous skeletal remains of organisms. Regarding this, as shown by V.I. Vernadskii in 1938, organisms play a dual role, by helping in the dissociation of the alumino-silicates and release of free silicic acid, which precipitate in the sediments.

A single opinion on the origin of silica in the rocks of the Bazhenovian formation is lacking. Some investigators emphasise the role of vulcanicity in the formation of silicification of the Bazhenovian reservoirs [15, 30]. The influence of pyroclastic materials in enriching the rocks of the Bazhenovian formation with silica has also been pointed out by R.S. Sakhibgareev and

R.A. Konysheva* [23]. A.V. Van [5] thinks vulcanicity exerted a decisive influence in the formation of the rocks of the Bazhenovian formation and suggests a model wherein vulcanic processes take on the role of supplying the source material, stimulating the development of planktonic microorganisms and transforming the OM into oil.

Our studies (1961–1986) have established that the major part of the silica of the Bazhenovian formation has a source of biogenic origin. At the initial stage of diagenetic transformation of the sediments, the quantity of impregnated waters diminished, resulting in the enrichment of pore waters inherited from the sediments saturated with biogenic silicic acid. Later, supersaturation of the solution and the acid precipitate silica resulted in silicification. This conclusion does not rule out the possibility of endogenic migration of silicic acid in the basin of the Bazhenovian period, favouring blooming of fauna with a siliceous framework.

Diagenetic silicification did not take place at one time. In the earliest stage, diagenetic silica filled the inner cavities of certain shells, which retained their original natural form (Fig. 18). Only because of the fact that silicification proceeded even in the flooded sediment, similar shells that were not silicified at the early stage of diagenesis showed silicification at variously deformed conditions during consolidation of the sediments.

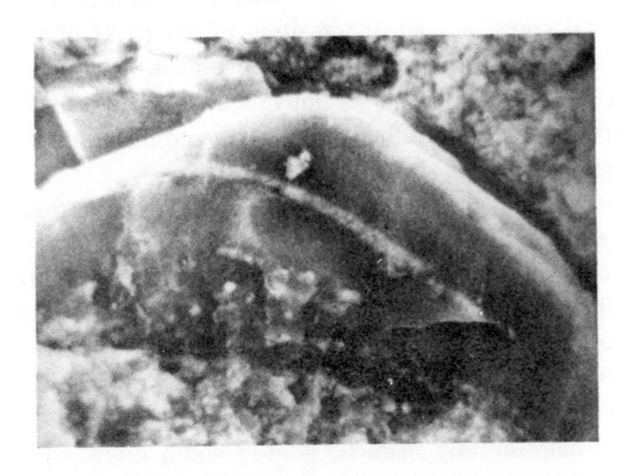

Fig. 18. Shells being silicified during early diagenesis. Salym area, well 118, depth 2787.1–2792.3 m; magnification 3000.

Diagenetic silicification of the Bazhenovian rocks led to the formation of a comparatively rigid framework within which the clayey matter consolidated with

*Given as R.A. Konysheva and R.S. Sakhibgareev in References—Language Editor.

less overburden stress. In the silicified parts, the clay minerals retained considerable plasticity and moisture. This is one of the reasons for the insignificant katagenetic transformation of clayey reservoirs of the Bazhenovian formation and the appearance of anomalous high plastic pressure (AHPP). There are also other points of view to explain the appearance of AHPP [39, 43].

The clayey reservoirs of the Bazhenovian formation differ from the overlying and underlying rocks as understood through X-ray structural and electronographic studies of their clayey components.

The rocks overlying and underlying the Bazhenovian formation are highly common in terms of their mineralogical composition. The principal component is hydromica, constituting up to 60% of the clay fraction. A sharp first basal reflection (1 nm) is typically prominent in diffractograms. A considerable admixture of mixed-layered minerals of the hydromica type, montmorillonite, is present in the overlying rocks to the extent of 35% and in the underlying rocks up to 60%. According to the data of electronography obtained by V.F. Chukrova, hydromica is present as a dioctahedral variety of modifications 1 M and 2 M_1 whose ratios vary in the samples. In some samples of the Achimovian formation, for example, their quantity is uniform. The parameters of the elementary unit cell vary within these limits: a–0.512 to 0.621 nm, b–0.887 to 0.902 nm.

In the rocks of the Achimovian and Abalakian formations of the Salym oil-field, the admixture of kaolinite is constant (around 5%). In wells with well-characterised cores, a gradual transition of the rocks of the Achimovian formation to those of the Bazhenovian is noticed. In this zone the reflections of hydromicas (degree of their crystallinity) are conspicuously degraded while the quantity of mixed-layered minerals increases and the admixture of kaolinite remains constant.

X-ray photographs of the rocks of the Bazhenovian formation, as shown by I.F. Metlova, above all differentiate the clay minerals according to their characteristic reflections (Fig. 19). The chief clay mineral of the Bazhenovian formation, as in the overlying and underlying deposits, is hydromica but it differs sharply in nature of reflection. If the basal reflection of the hydromica (1 nm) of the Achimovian and Abalakian formation is about 15 cm, then that of the Bazhenovian varies from 0.5 to 2 cm, that is, nearly ten times less. While the reflection is of a halo-forming character with its width almost equal to its height, it is sharply expressed as a loop in the region of small angles. Such a character of reflection of hydromica might indicate a significant admixture of a swelling component. Still, treatment of the sample with glycerine did not reveal the presence of hydromica and the X-ray photograph registered no change. According to the data from electronographic studies, this mica, like mineral hydromica, is montmorillonite (1 M) with lowly developed structures, the ratio of layers 2:1 and the following parameters of the elementary unit cell: a–0.512 to 0.514 nm, b–0.886 to 0.890 nm.

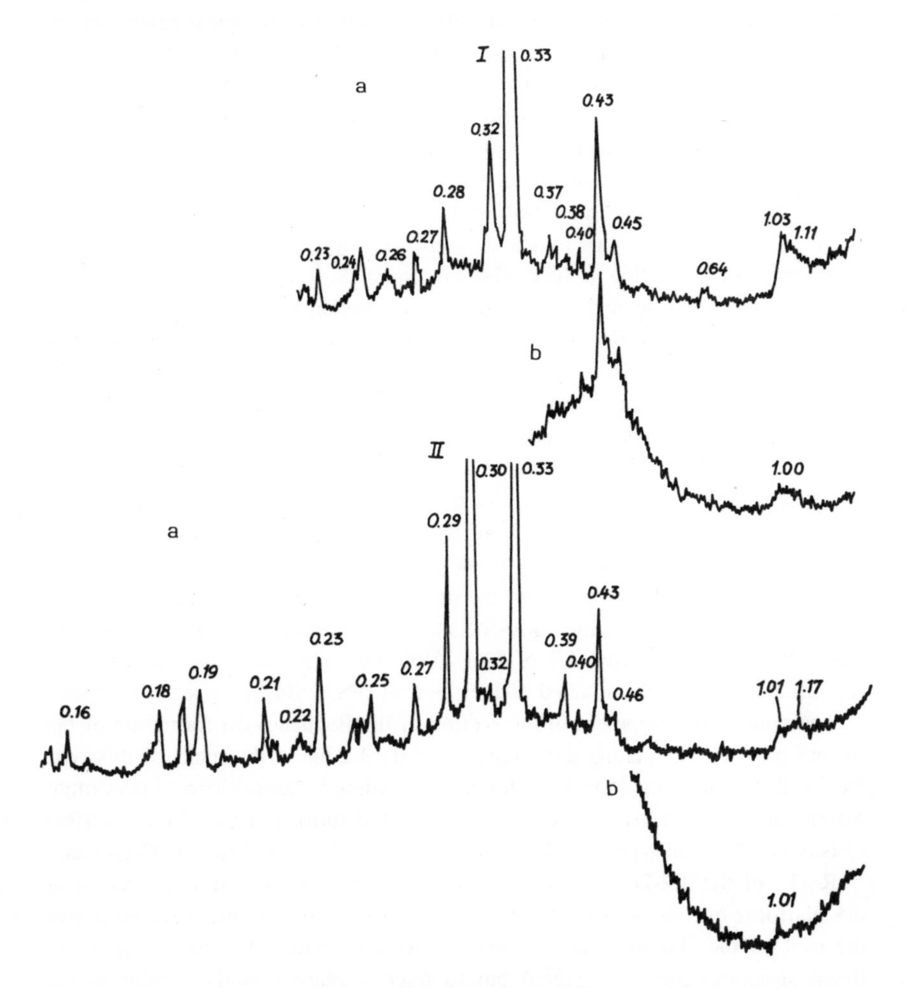

Fig. 19. Diffractograms of rocks of the Bazhenovian formation, Salym oil-field (a—original sample; b—sample saturated with glycerine):

I–well 135, depth 2983–2990 m;
II–well 118, depth 2763–2366.7 m [*sic*].

To distinguish the types of rocks constituting the section of the Bazhenovian formation, several workers have used different characteristics: T.V. Dorofeeva and her co-authors [21] used mineralogical-chemical and petrographic; I.I. Pluman, P.G. Demina, F.G. Gurari and N.I. Matvienko [15, 34] used radioactivity; and O.G. Zaripov, V.P. Sonich, I.N. Ushatinskii and others [9, 15, 30, 42] have used enrichment of OM and the degree of its bituminisation. Many workers, however, consider the apparent electrical resistivity an important criterion for classifying the rocks of the Bazhenovian formation.

E.A. Gaideburova [8] distinguished four types and eight subtypes of cross-sections of the Bazhenovian formation on the basis of apparent resistivity (AR) values and the state of maximum radioactivity. While identifying the subtypes, the form of the curves of AR and gamma-ray logging (GL), the number and position of their maxima in relation to the formation, the nature of fall (low) of curves AR and GL and other indicators were taken into consideration. Analysing the distribution according to the field of commercial oil flows in the Bazhenovian formation, E.A. Gaideburova came to the conclusion that oil prospects might be considered in territories with deposits possessing geophysical characteristics indicating sections of the fourth (Salymian) and the third type. In the latter, of the five subtypes distinguished as commercially exploitable, E.A. Gaideburova considers only two quite promising–the Malorechenian and the Kochevian.

The section of the Bazhenovian formation of the *first (transitional) type* is developed over the limited territory of the southern part of the western Siberian platform. Eastwards, in the region of Krylov oil-field, rocks of this type thin out. The section is represented by weakly bituminised rocks with intercalations of non-bituminised varieties. From the deposits of the exposed sequence of the Mar'yanovian formation, the rocks of the Bazhenovian formation of the first type differ by an increased AR of 5 to 25 ohm·m. The maximum radioactivity is confined to the rocks situated above the roof of the Bazhenovian formation.

The zone of development of the section of the Bazhenovian formation of *the second type* stretches along the boundary of the basin. In the rocks of this type the AR fluctuates from 25 to 125 ohm·m. According to the position of maximum gamma activity and the degree of division of the formation based on electrical resistivity, three subtypes are distinguishable: Tai-Dassian, Bratian, Tagrinian.

Rocks of the Tai-Dassian subtype are developed in the southern regions of the platform over a limited territory. The maxima of GL are situated above the roof of the Bazhenovian formation. Sections with one-, two- and three-tiered structures are encountered but to trace a sharp boundary between the members with various resistance is impossible. In zones of reduced thickness the Bazhenovian formation lies over the rocks of the Tyumenian formation. Deposits of the Geogrievian formation are practically absent at these places.

Deposits of the Bratian subtype of the Bazhenovian formation extend as a narrow belt along the zone of the Tai-Dassian subtype, from which they differ only in the location of the maximum gamma activity, belonging to the upper part of the formation.

Rocks of the Tagrinian subtype are encountered in two different regions of the western Siberian platform. Maxima of gamma activity are found in different parts of the formation and are not noticed beyond its limits. Rocks are characterised by considerable thickness (70 m) and alternating highly bituminous and non-bituminous rocks.

The section of the Bazhenovian formation of *the third type* is distributed over the major part of the Central Pri-Ob′ territory. AR values range from 125 to 625 ohm·m. Maxima of gamma activity are encountered in a number of parts of the formation. High AR values and gamma activity are not always coincident. Based on these indicators, five subtypes have been identified in the third type: Tai-Tymian, Festival'nian, Malorechenian, Kochevian and Yuganian.

The description of the deposits of the Bazhenovian formation in the sections of the Tai-Tymian, Festival'nian and Yuganian subtypes seems to negate their prospects for oil potential.

Deposits of the Malorechenian subtype are spread over the greater part of the Yugan depression on the western slope of the Kaimysov dome, Lar'-Egan arch and the row of fields of the Lower Vartov and Surgut anticlines. The thickness of the formation increases from the Lar'-Egan arch (6–8 m) through the Lower Vartov anticline (10–15 m) towards the Southern Surgut oil-field (26 m). The maximum gamma activity is noticed in the central part of the formation. In the zone of development of the rocks of the third type are included two local zones of the section of the second type–in the central part of the Surgut anticline (Mil'ton, Minchimkin and Yur'ev fields) and in the region of the Ur'ev oil-field.

Rocks of the Kochevian subtype developed in the Surgut anticline (Kochev, Tavlin, Kogolym, Savui and other fields). Each member identified in the subtype is characterised by a peak in increased gamma activity values. The section of the Kochevian subtype possesses a complex structure with distinct changes in thickness with the exposure of individual layers and members.

The section of the *fourth (Salymian) type* is a typical standard one with country rocks enclosing commercial pools of oil. Increased AR values reach 4500 ohm·m. In a number of cases in the rocks of this type, negative anomalies are observed on the potential self-polarisation (PS) curve. The maximum values of gamma activity are confined to the upper part of the formation. Based on the curve of gamma activity in this type, seven members have been identified in this section [43]. The thickness of the Bazhenovian formation in the Salym region varies from 35 to 50 m.

Detailed division of the section of the Bazhenovian formation into members according to GL data (see Fig. 18), and the natural radioactivity of the rocks depending on the degree of bituminisation, was determined by V.V. Khabarov, O.M. Nelepchenko and T.V. Pervukhina. The section is divided into six lithostratigraphic members (from top to bottom): C_1, P_1, C_2, P_2, C_3, P_3 (see Fig. 17). Each pair of members C and P comprises a sedimentation complex of geological bodies, which possesses a set of stable characteristics. For these complexes a sharp variation in the distribution of the majority of the elements has been established. A rhythmic pattern of changes in the major geophysical, physical and material characteristics and the uranium and C_{org} contents in them has likewise been confirmed.

Studies carried out by us on the rocks of the Bazhenovian formation from the Berezov-Shaim region showed that the rocks developed in the Salym cupola and its setting considerably differ from the rocks developed in the western, southern and eastern borders of the basin. Deposits constituting the section of the Salym type in fact serve as a standard for three component clayey reservoirs of the Bazhenovian type, by means of which it is possible to carry out a comparative estimate of reservoirs of analogous types in other regions.

The rocks of the Bazhenovian formation developed in the Salym oil-field and adjacent prospecting fields close to the south-west boundary of the Surgut anticline at the zone of its interlink with the Mansii syncline have been studied thoroughly. Here they lie at the absolute elevations of 2697–2848 m at the formational temperature of 120–128°C and AHPP (with excess of hydrostatic pressure at 14–20 MPa).

Microscopic studies of the rocks reveal characteristic textures in them at the meso- and microlevels, formed during the geochemical interaction of the rock-forming components with one another (clay minerals, OM, silica) in the process of sediment formation.

OM of the second type played a dominant role in the formation of mesotextures. It is distributed in the form of lamellae and as lenticular clots which gave rise to the appearance of parallel-lamellar, mesh or reticulate and lenticular-lamellar mesotextures (Fig. 20a). The presence of lamellar or layered textures with OM of the second type led to anisotropy of the strength and filtration characteristics of the rocks. The compressive strength of the rocks perpendicular to the bedding is 5.2 and 5.5 MPa and parallel to the bedding, 3.8 MPa (Salym field, well 32). The permeability of the rocks along the bedding is twice higher than that perpendicular to the bedding (corresponding to 10^{-7}–10^{-9} and 10^{-9}–10^{-11} μm^2). These data of E.P. Efremov and others accord with our data obtained during experimental modelling of the process of migration. Some role in the formation of mesotextures was also played by the terrigenous silty and coarse pelitic admixtures.

Microtextures formed during the distribution of microblocks and microaggregates of clay minerals around centres of stresses which, in the rocks of the Bazhenovian formation, are the terrigenous and carbonate minerals, silica concretions, OM of the first type and other components of the rocks stronger than the clay minerals. Textural heterogeneity at the microlevel is especially significant. It appears in the blocks where the clay minerals and silica are impregnated with OM of the third type (Fig. 20b). The OM adsorbed by the minerals hydrolysed their surface and significantly lowered the adsorption capacity or, in other words, increased the reservoir potential.

The three chief rock-forming components of the Bazhenovian formation

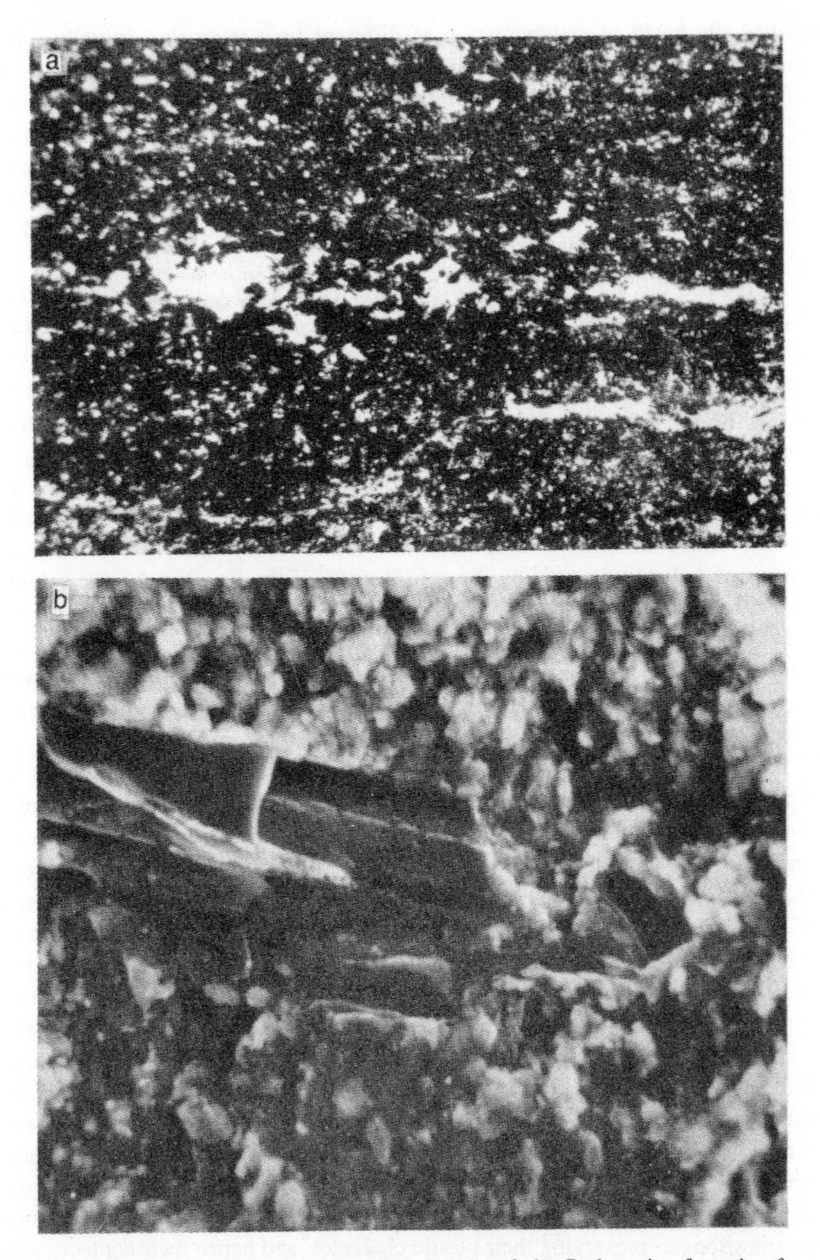

Fig. 20. Mesotexture (a) and microtexture (b) of rocks of the Bazhenovian formation from Salym oil-field (well 32); a—depth 2749.42–2764.07 m; magn. 18, Nichols II; b—depth 2764.42–2770.32 m; magn. 1000.

reached the basin of sedimentation at different time intervals. Initially, clay minerals and the allocthonous part of the OM reached the basin from the fields of mobilisation of the material. Because of the rapid rate of ion-exchange reactions, the clay minerals and OM entered into interaction even as they began to pass into the sediments and the exchangeable ions of clay minerals converted into organic. This process was accompanied by primary hydrophobisation of the surface of the clay minerals. Then the autocthonous part of the OM formed in the basin.

As pointed out earlier, the process of precipitation of silica and silicification of the Bazhenovian sediments is related to diagenesis. It has been established that silica did not fall out of the solution immediately in a crystalline form, but formed colloidal flocculents or clots. Colloidal silica enveloped aggregates of clay minerals both with and without the adsorbed OM. Thus, highly complex products were formed from clay minerals, OM and silica.

A high content of OM in the rocks and pore waters is characteristic of the Bazhenovian period. In the rocks of the Salym oil-field, C_{org} varies from 15–23%. Hence in diagenesis at the initial stage of recrystallisation of the components of sediments providing the constituents for the future rocks, the soluble component parts of the newly formed OM were adsorbed by the clay minerals and collomorphic silica. The silica enclosed scales of clay minerals or localised in the form of concretions of various shapes hydrolysing the surface of the clayey microblocks without siliceous pockets, clusters of silica and complex systems composed of clay minerals, OM and silica in different combinations.

The increased formational temperature of the Bazhenovian formation left its imprint on the sediments through various changes, including that of transformation of the OM contained in them. Under the influence of increased temperature that part of the OM adsorbed by the rocks transformed into hydrocarbons (HC). According to the data of I.A. Yurkevich, the proportions of syngenetic HC present in the OM of the rocks of the Bazhenovian formation studied from well 49 amounted to 12 to 15% of all the hydrocarbon components contained in these rocks. Simultaneous with the thermocatalytic action, the thermal transformation of OM of the first and the second type also took place. But as the ratio of clay minerals to OM in the rocks of the Bazhenovian formation is not high (2:1, rarely 3:1) and as in the OM a significant content of components of the first and the second type is observed, then as a result of the thermocatalytic and thermal transformation of OM a large quantity of heavy hydrogen-poor products remained in the rocks, irreversibly adsorbed by the minerals according to the laws of chemical and physical adsorption. Hydrophobised blocks of the rock became stronger, preserving their plastic character, and hence their contacts with other parts of the rocks were less strong than the contacts of the two parts without the OM. Hydrophobisation of the surface of minerals predetermined the possibility of these rocks becoming effective reservoirs.

Let us examine the patterns of transformation of OM into petroliferous hydrocarbons (HC) established by us as one of the current points of view on the formation of commercial oil-pools in the reservoirs of the Bazhenovian formation of the Salym oil-field. According to the opinion of the proponents of this view [21], the clay comprising the Bazhenovian formation was originally montmorillonite. As a result of the transformation of OM, it transformed into hydromica, which led to the disintegration of the rock due to desorption of the newly formed HC by the surface of the 'hydromicaised' montmorillonite and the formation of oil-pools in the petroliferous rock.

The question of suppression of transformation of montmorillonite into hydromica was studied by the author from the point of view of balance of chemical elements [19] and by M.F. Sokolova from the point of view of differences in the dimensions of lattice charges of montmorillonite and hydromica [37] and hence is not discussed here. We only believe in the possibility of disintegration of the original montmorillonite minerals during the OM-HC transformation. An important characteristic of clay minerals, particularly of montmorillonite, is their high ion-exchange capacity. The larger the size of the organic ion and the higher its charge, the stronger its adsorption on the surface of the mineral, when both Coulomb bonds and Van der Waals forces are at work, interacting with variously charged particles. However, the larger the ion, the greater the Van der Waals forces.

It is also well known that the ion-exchange reactions are characterised by high speed. Hence the clay minerals already on their way through formational waters in the sediments attain equilibrium with the cations of the sea water from which, in the first instance, the organic ions were adsorbed. In diagenesis and katagenesis, as a result of thermolysis and thermocatalysis, newly formed hydrocarbons (HC) were generated, which were adsorbed by freely exchangeable positions of clay minerals.

Studies by A.D. Petrov, N.D. Zelineskii, A.V. Frost, A.I. Bogomolov, N.B. Vassoevich, S.G. Neruchev, A.A. Petrov, L.A. Gulyaeva, V.A. Uspenskii and others have shown that the higher the ratio of mineral part to organic, the deeper the transformation of OM and the less the compaction of the products on the surfaces of the minerals. It is interesting to note that this pattern was established by N.A. Eremenko in 1940, while working on the petroliferous deposits of the Caucasus.

The minor preponderance of the mineral part over the organic in the rocks of the Bazhenovian formation of the Salym oil-field led to the appearance of numerous slags permanently or irreversibly adsorbed by the minerals in the process of transformation of OM. Montmorillonite, because of the exceedingly high dispersion of the particles, favours the most intense transformation of OM but, concomitantly, as shown by our experiments, it does not readily release whatever is adsorbed further under severe conditions of alcohol-benzene extraction. Further, montmorillonite with the adsorbed OM in the exchangeable positions does not change further into minerals, which according to geochemical and energy (thermodynamic) data, could, perhaps, undergo further transformation.

Thus, if the original composition of the clayey part of the rocks was montmorillonite, then their hydrophobisation was still more intensive and no commercial accumulations of oil could ever have been formed. The disintegration of the rocks did not take place because OM adsorbed by montmorillonite served as a peculiar groundmass joining together the rocks because of which their screening effect was increased and not their reservoir property. We further add that even then a small quantity of autocthonous bitumens, which formed in the rocks of the Bazhenovian formation from typical OM, became dissociated from the rocks in the case of the montmorillonite composition of their clayey part, and was left over on the adsorbing surface of the minerals, that is, could not contribute its own amount to the oil saturation of the formation. This proceeded further and, in such a case, if the presence in the rocks of collomorphic silica is not taken into consideration, the high adsorption capacity is understandable.

While studying the structures of pore space and its quantitative characteristic obtained through the method of R.A. Konysheva and A.P. Roznikova [22], it was established that for the Bazhenovian rocks the specific surface which attests to the sinuous nature of the pores and their small dimension, acquires special significance. The influence of the textural aspect of the pores on their configuration is estimated by the coefficient of orientation, which is expressed as the ratio of extension of the pores in two mutually perpendicular directions. According to the data on wells with low productivity, for lamellar textures the coefficient of orientation varies from 1.100 to 1.135 and in massive ones does not exceed 1.040. Data for the highly productive wells are presented in Table 4.

Table 4: Parameters of meso- and microtextures

Parameters	Mesotextures	Microtextures*
Coefficient of orientation	1.446	2.229–1.334 (1.623)
Ratio of length of zone (mm) in terms of miscalculated area (1 mm^2)	5.25	63.40–85.17 (70.75)
Total length of contacts (mm)	28.41	0.792–1.065 (0.884)

*Figures in parentheses–mean arithmetic values.

The open porosity of the Bazhenovian rocks determined by the method of transform over Quantimet-720 varies from 4.5 to 9.5%. These data accord with the results obtained by V.M. Dobrynin and V.G. Martynov by the computational method.

The mass determinations of porosity carried out in the laboratory of plastic physics VNIGRI showed that the open porosity of the rocks of the Bazhenovian formation changes from 2 to 16%. To explain the variation in sizes of open porosity, the straight and conical methods of qualitative and quantitative estimates of the capacity of the sedimentary formations were employed [21]. The conclusion drawn was that the fractures in the Bazhenovian formation cannot provide significant value for estimation of capacity. The basis for the capacity of a reservoir is the cavities of the matrix developed, which are connected by hydromicaised montmorillonite. This process facilitated primarily the formation of horizontal fracturing and favoured connection of the cavities in blocks of rocks and the appearance of channels of filtration in the layered formation.

According to the data of the Central Laboratory of Glavtyumen Geology, the rocks of the reservoir layer Yu_0 are characterised by comparatively not high values of open porosity (average, 3 to 8%). The effective porosity is retained only because of the fracturing. A similar behaviour in the nature of the capacity of the layer Yu_0 has been presented by the collaborative work of ZapSibNIGNI and SibNIINP. To determine the reservoir properties of the Yu_0 layer, different geophysical methods were employed: LLS (lateral logsounding) and LL (lateral logging) (ZapSibNIGNI), impulse neutron-neutron logging and thermometry up to and post-injection of salt (VNIIYAGG, ZapSibNIGNI).

The permeability intervals in the layer Yu_0 were not delineated using standard industrial-geophysical practices. V.V. Khabarov, O.M. Nelepchenko and T.V. Pervukhina conducted geophysical measurements in a series of wells according to the method of double solutions. However, even by this method it was not possible to establish the critical meaning of geophysical parameters for division of permeable and impermeable intervals of the section. A distorted picture was obtained due to the considerable electrical heterogeneity of the bituminous rocks.

Under conditions of frequently alternating intercalations of 0.4–2 m thickness with apparent electrical resistivity reaching 10 to 3,000 ohm·m and above, the lateral logging (LL) method is the most effective.

The structure of the pore spaces of the Bazhenovian rocks was studied in stained thin sections at VNIGNI [6]. The rocks were soaked in formaldehyde-resorcin-resin with formalin as the hardener and rodamin dye, according to the method described by L.A. Kotseruby. A study of the slides showed that the sizes of the pores were almost the same in all the samples–from 0.02 to 0.06 mm, rarely going up to 0.10 mm. The pores were well connected among themselves and situated along the layers. In such parts, the porosity was 25–28%. Where the connection between the pores was poor, the porosity estimated from the thin sections amounted to as little as 5%.

As seen from the data presented above, the various methods used in the study of the reservoir rocks did not yield unambiguous results. The ambiguity of interpretation of the data on the reservoir parameters denies an acceptable explanation of the output values obtained in the experimental wells; 0.06 to 300 m^3/day (Fig. 21). According to the data provided by N.A. Krylov, B.V. Kornev and M.I. Kozlova [25], a certain pattern is discernible in the oil yield from the different fields in the Salym petroliferous belt. In the folded part of the structure lies the zone of high oil output, with yields decreasing as one proceeds away from it; thus, initially the discharge was 10–15 m^3/day, reducing to 1–10 and farther away to less than 1 m^3/day, and finally 'dry' wells appeared. In the eastern part of the structure, a second zone of increased production of oil appeared. If we consider the data on the laboratory determinations of the physical constants of the reservoirs of the Bazhenovian formation (porosity, %: general–7.33, open–4.29, fractured–0.09; permeability, μm^2: fractured–0.001, for gas–0.018), then the lack of correlation between the physical properties of the reservoir rocks and the oil yields obtained from them is obvious.

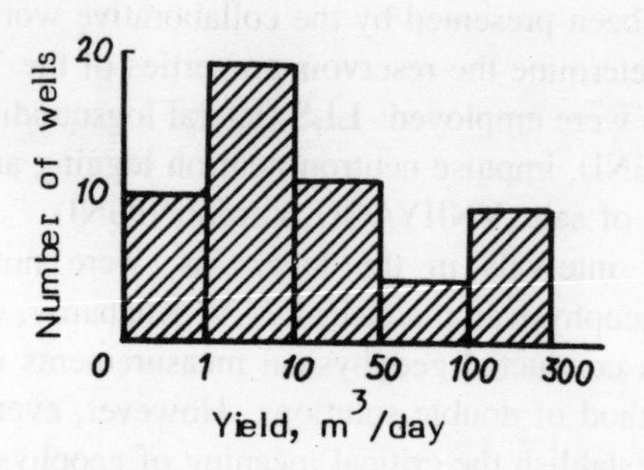

Fig. 21. Histogram of oil yields in the Salym oil-field [24].

This lack of correspondence between the physical constants of the reservoirs, determined by different methods, and the actual oil yields can be explained as due to the fact that not only the pores, as illustrated in Table 1, but also the zones of textural heterogeneity (see Table 4) influence the capability of such specific reservoirs as the clayey reservoir of the Bazhenovian formation. It is precisely these zones which play a decisive role in the formation of such infiltration characteristics of rocks.

In the photomicrographs of the reservoirs of the Bazhenovian formation obtained over SEM (Fig. 22), the different microtextures both along and perpendicular to the bedding are visible. Canals 1.5 to 5 μm wide are well seen, constituting the separation of the parts with various microtextures during tectonic stresses related to the formation of the oil and gas-pools in these rocks. During this, as shown in Table 4, it can be seen that in the direction perpendicular to the bedding, the canals are 1.6 times larger than in the direction parallel to the bedding. It is worthwhile recollecting once again that these are not the fractures but the disintegrated weakened zone under extreme conditions because in the Bazhenovian rocks, in spite of the presence of a hard siliceous framework, plasticity is sufficiently high (coefficient of plasticity 1.6–2.3). Hence, their tendency to form fractures is also not high (less than 10 units in 1 m, according to the data of IG and RGI).

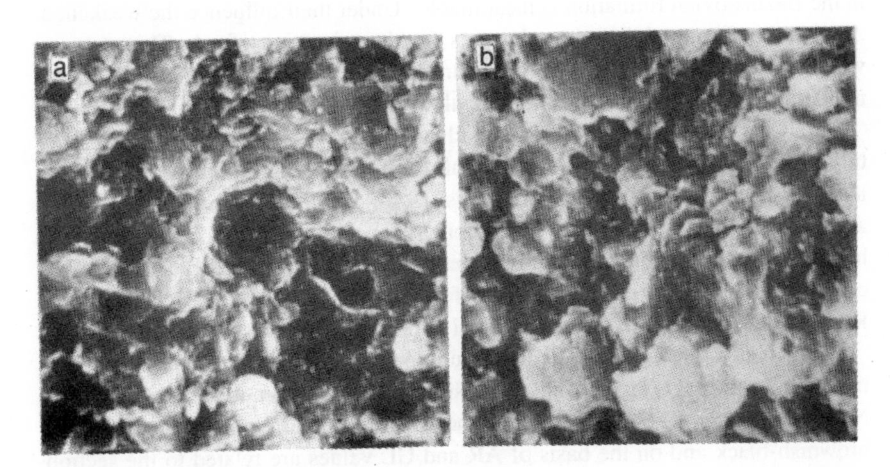

Fig. 22. Microtexture of rocks of the Bazhenovian formation parallel to bedding (a) and perpendicular to it (b) Salym oil-field, well 32, depth 2770.32–2775.31 m; magn. 1200.

The disintegration of the weakened zones and the formation of the reservoirs in the Bazhenovian and other clayey rocks constituted a process in simultaneous operation at the moment of migration and accumulation of oil in the rocks.

For the Jurassic period two major stages of intense tectonic activity are characteristic of the Salym region. The first stage was marked by the formation of Jurassic deposits over the strongly dissected erosional surface of the heterogeneous basement. G.M. Taruts and E.A. Gaideburova [41], based on the results of complex interpretation of geological, geophysical, lithological and analytical studies of the rocks of the Bazhenovian formation of the Salym region, established a sharp block structure of the pre-Mesozoic basement. The blocks possess diverse dimensions, forms and trends as a result of a system of faults of varied orientations and different ages. The same authors showed that the structure of the Salym region was determined by its situation in the regional plan at the junction of two large blocks of the earth's crust–the ancient Uvat-Khanty-Mansii central massif on the west and structures of the young Hercynian fold system (Salymian) on the east. The latter trends north-westwards from the south and takes a sharp turn north-east at the zone of junction with the central massif. In this zone the large shear of the Hercynian structures in the north-east developed along a deep-seated fault during the Triassic. At that time the principal system of the local blocks of the basement of the Salym region formed. The late Jurassic stage was characterised by the energetic growth of the synsedimentary structures. Disjunctive faults are widely developed and traceable not only in the basement, but even into the sedimentary cover.

The significance of the tectonic stresses during the formation of the reservoir in the Bazhenovian formation is inestimable. Under their influence the weakened zones were disrupted, opening channels for the migration of oil. This process was intensified under the influence of increased temperature, which facilitated movement of the vapour-gas solutions along the faults.

Analysis of the materials from the clayey rocks of the Bazhenovian formation of the Salym oil-field has been helpful in formulating the major mineralogical, geochemical and textural prerequisites for the formation of clayey rocks as potential reservoirs: textural heterogeneity, specific material composition and hydrophobisation of the surface of minerals by OM and silica. Based on these criteria, the rocks of the Bazhenovian formation from the wells opened in the Krasnolenin dome in the Khanty-Mansii and Nyurol' depressions were tested.

The Khanty-Mansii depression or trough presents its own characteristic structure; the main stage of sagging or subsidence spread over the Cretaceous and Neogene periods. The rocks of the Bazhenovian formation in this depression are brownish-black and on the basis of AR and GL values are related to the section of the second type [8].

Studies of thin sections of the rocks of the Bazhenovian formation from wells opened in the Khanty-Mansii depression reveal the almost complete absence of OM of the third type and also a change in colour of OM of the second type. If in the Salym oil-field OM of the second type varies in colour from dark brown to distinct black, then in the wells of Khanty-Mansii depression this OM is reddish-brown. It is in the form of very thin (0.02–0.01 mm) discontinuous bands or patches of OM of the first type of larger size than in the Salym field. These bands of OM form indistinctly layered mesotextures; the fine granular silty part (up to 5% of the rock) is unevenly distributed (Fig. 23a).

The mixed-layered minerals of the hydromica type–montmorillonite, hydromica, kaolinite–constitute the rock-forming clay minerals. A little chlorite is present and its content varies from individual grains in the lower part of the formation to small quantities in the upper, as deciphered through diffractograms. The dispersion of clayey components in the rocks diminishes from the bottom to the top along the section. As in the Salym field, a considerable amount of pyrite is present in the rocks of the Khanty-Mansii depression. It completely replaces OM of the first type and is also formed as a product of transformation of OM of the second type. Pyrite is found together with OM and also at some distance from it. Bituminous components formed during the transformation of OM of the second type and, remaining adsorbed by the clay minerals, played some role in producing textures at the microlevel (Fig. 23b). Microtextures are primarily axial but in the complex layered minerals they are less orderly than in kaolinite and particularly hydromica.

The section of the Bazhenovian formation of the Krasnolenin dome belongs to the same type as that of the Khanty-Mansii depression, as shown by electrical characteristics. At the north-western pericline of the dome in the region of Talin field, the rocks of the Bazhenovian formation present as a series of dark brown to black clays in the upper part of the section and dark brown at the lower. The clays are indistinctly layered because of the distribution of very thin layers and lenses of OM that are either colourless or differ in degree of coloration (Fig. 24). Elongate patches of OM of the first and second type are distributed along the bedding. Many large patches of spore fragments of reddish hue in particular provide reddish tones to the clays. In some layers the clay minerals are scattered with small crystals of carbonate, more often siderite. In well no. 1 of Talin field, intercalated dolomitised limestone with calcitic faunal remains is encountered. Individual inclusions of OM of the first type and patches of reddish spore fragments are present in these rocks. The clay minerals present are hydromica and mixed-layered minerals of the hydromica type–montmorillonite and chlorite. Kaolinite is present in minor amounts.

In the Pal'yanov oil-field (south-east pericline of the Krasnolenin dome) the rocks of the Bazhenovian formation are analogous to those exposed in the Talin oil-field. The difference, however, lies in the fact that the finest weathered scales of hydrobiotite are observed along the bedding. Amongst the clay minerals in such samples, hydromica predominates, followed by the mixed-layered hydromicas–montmorillonite, chlorite and kaolinite–which here occur more in the higher parts of the section.

The rocks of the Bazhenovian formation exposed in the wells on the Krasnolenin dome and in the Khanty-Mansii depression differ from the rocks of the Bazhenovian formation of the Salym oil-fields, above all in their type of OM and its quantity (maximum 6%), and also in their silicic acid content (up to 8%). In these clays there is very little OM, normally adsorbed by the clay minerals, and hence hydrophobisation of clay minerals is by no means reflected in the character and intensity of the first basal reflection of hydromica in the diffractograms. Only in the lower part of the range, 2416–2428 m of well 51, in the Pal'yanov field, are the reflections of the hydrated hydromica (1.02 nm), for example, 5 times less in intensity than the reflections from a sample without adsorbed OM, and reveal a small loop in the region of smaller angles (Fig. 25, I).

The Nyurol' depression is situated at the centre of the central tectonic field and extends in a northerly direction. The deposits of the intervening complex filling this depression are broken up by a network of multidirectional faults, forming a system of uplifted and subsident blocks. Within the limits of the subsident blocks, depressions and megadepressions formed and were filled by

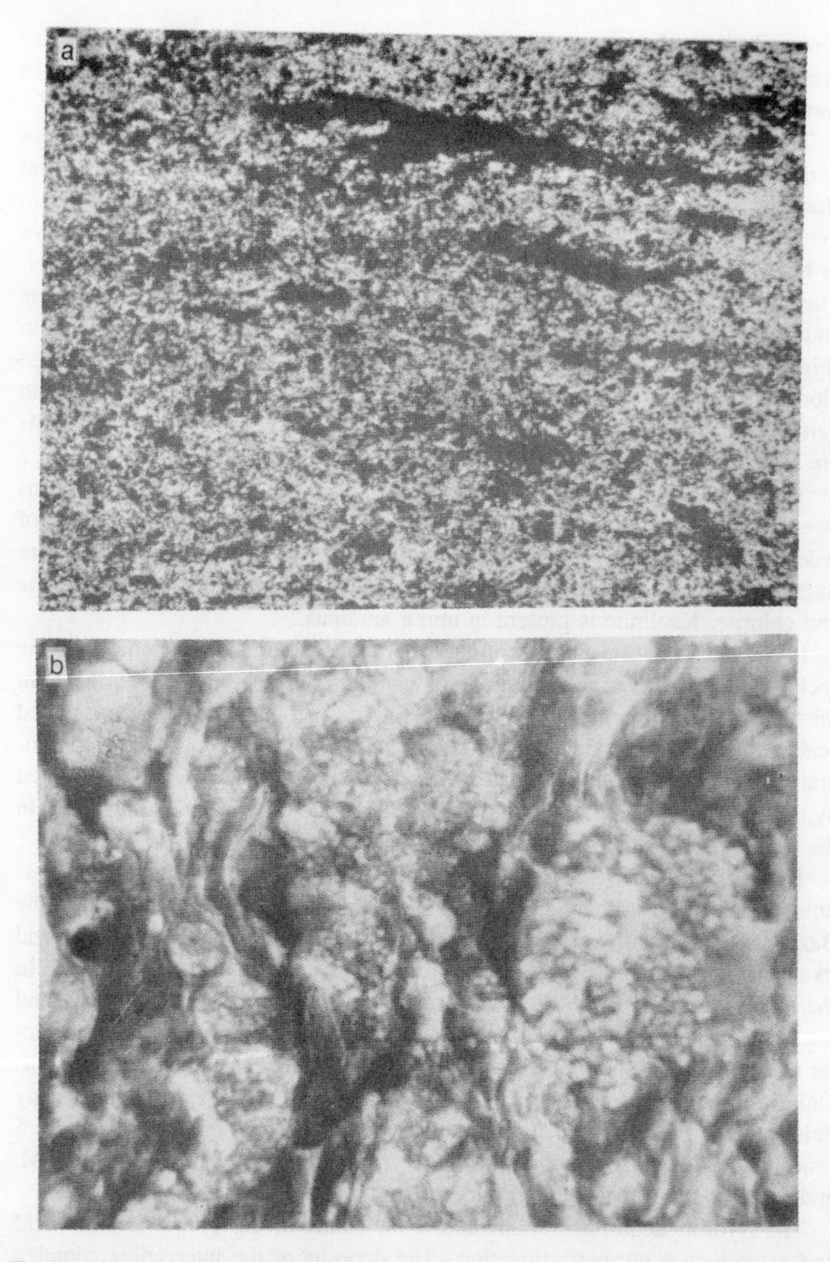

Fig. 23. Mesotexture (a—magnification 15, Nichols II) and microtexture (b—magnification 1000) of rocks of the Bazhenovian formation of the Khanty-Mansii depression (well 3, depth 2788-2794 m).

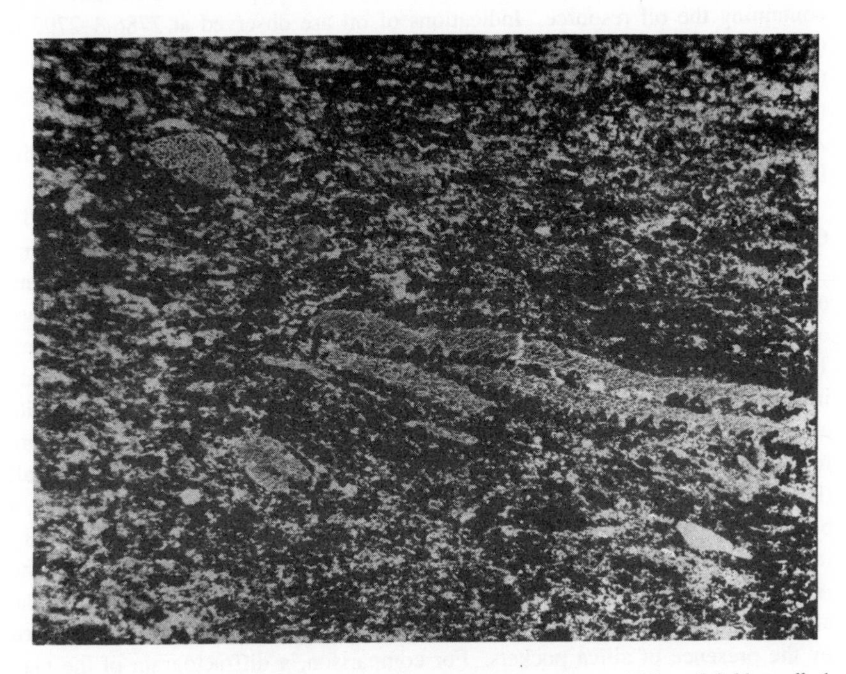

Fig. 24. Mesotexture of rocks of the Bazhenovian formation. Talin oil-field, well 1, depth 2407-2416.4 m; magn. 15, Nichols II.

a thick series of deposits of the intervening complex (up to 4000 m). The uplifted blocks form ridges, massives and benches in which the thickness of the intervening complex is reduced to 500 m and in some parts the deposit thins out completely [9].

The Nyurol' depression occupies a distinctly larger area in the pre-Jurassic deposits than in the sedimentary cover. Borehole data have proved a partially inherited tectonic signature–local structures of the sedimentary cover repeat only the morphology of the surface of the pre-Jurassic deposits and cannot be correlated with the character of their disposition. Constant inherited development is observed only in the region of Mezhov ledge and Kalgach ridge, which in the cover correspond to the Mezhov anticline and Kalgach uplift.

In the Nyurol' depression the oil flow from the Bazhenovian formation is obtained in the Eastern Moiseev oil-field. In well 1, drilled over the domal part of an analogous uplift, the Bazhenovian formation is found at depths of 2767–2792 m. The rocks are dark brown, bituminous, compact and micaceous, with inclusions of coaly detritus, seams of calcite and faunal remains. The thickness here is 25 m. At the range of 2773.4–2777.8 the core (1.7 m) shows dense bituminous rocks which, at the end of the layer, change over to a fractured type containing the oil resource. Indications of oil are observed at 2786.8–2792 m (the lower part of the formation). The AR values at the Eastern Mioseev structure reach 275 ohm·m, the formational temperature is 104°C and the formational pressure, 55.6 MPa. At well 2, this same structure occurs at a depth of 2762–2788 m and an oil yield of 2.4 m^3/day is obtained from the rocks of the Bazhenovian formation [15].

We studied core samples of the Bazhenovian deposits from the Igol' (well 11), Chertalin (well 1) and Sel'veikin (well 3) oil-fields of the Nyurol' depression.

In the Igol' well, the Bazhenovian rocks reveal hydromica with an admixture of mixed-layered minerals and kaolinite. They are indistinctly layered because of extended seams 0.02 mm × 0.30 mm, of darker hues than the groundmass. The colouring of the seams is facilitated mainly by such a microtexture and only in some places by a large degree of adsorption of OM. Concretions of calcite are distributed in a nearly layered fashion but with large scattering. There are many reddish spore fragments of varying sizes in the rock, from very small (0.02 × 0.04 mm) to very large (1.4 × 0.5 mm), whose presence favours a high content of C_{org} (8.34%), which is quite typical of these rocks. Some very large fragments are mineralised (colouring). A considerable amount of pyrite is seen along with finely dispersed OM of the first type. Reflections of the clay minerals are of very low intensity (Fig. 25, II), which can be explained by the presence of silica pockets. For comparison, a diffractogram of the clay sample from the Georgievian formation, directly underlying the Bazhenovian, is given in Fig. 26. The integrated intensity of hydromica from the sample of the Georgievian formation, for example, is 12 times higher than in the sample from the Bazhenovian formation. The block microtextures lack distinct orientation (Fig. 27).

The Igol' structure is situated in the same meridian as the Eastern Mioseev structure but distinctly southwards, bordering the zone which A.E. Kontorovich and others [9] consider unproductive. Although no final conclusion has been reached, for the present it may be said that the absence of one of the principal factors, in particular adsorbed OM, hydrophobising the contacts of microblocks and microaggregates of clay minerals, does not permit evaluation of this structure along the Bazhenovian formation as a highly potential one, even though the intersection of a fault through this structure might help in the disruption of the weakened zones with siliceous pockets, which are considered favourable sites of accumulation of oil. Further studies are necessary therefore.

In well 1 of the Chertalin oil-field the Bazhenovian formation is composed of rocks showing indistinct, parallel or lens-like layering with a discontinuous layer of dolomite in the middle of the sample from the depth range of 2769–2780.5 m. A not very clear layering demarcates OM of the first and the second type distributed along the bedding. The hydrolysed components of OM of the second

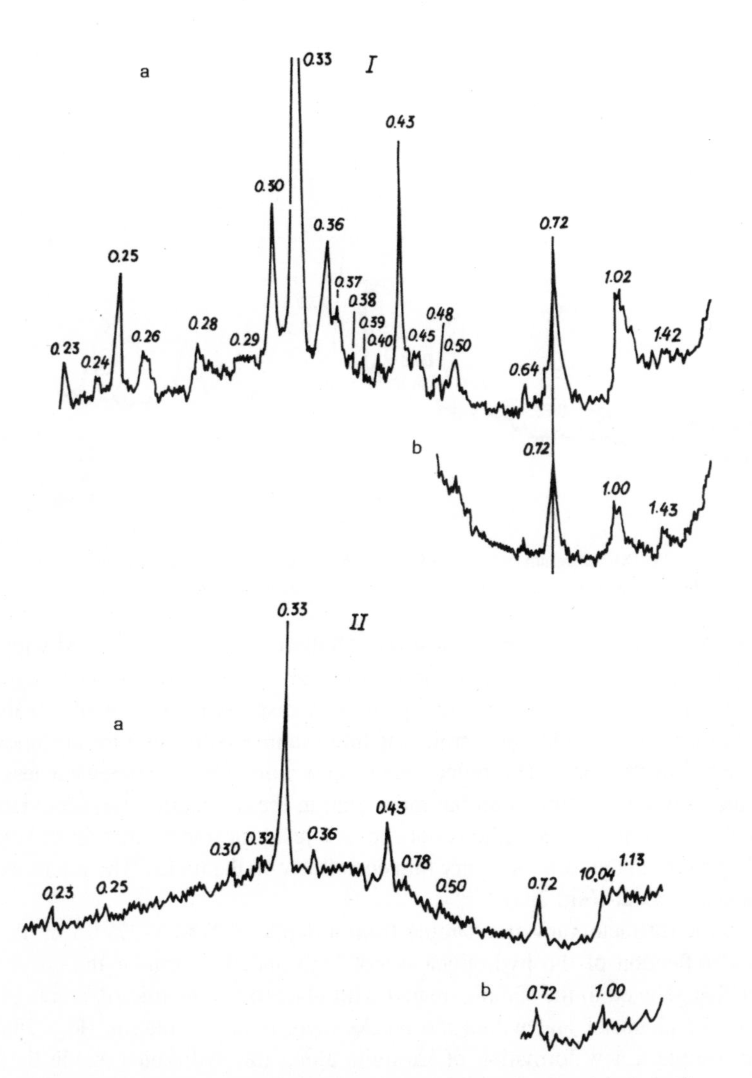

Fig. 25. Diffractograms of rocks of the Bazhenovian formation (a–original sample; b–sample soaked in glycerine).
I–Pal'yanov oil-field, well 51, depth 2590.2–2600.2 m; II–Igol' oil-field, well 11, depth 2792.1–2799.1 m.

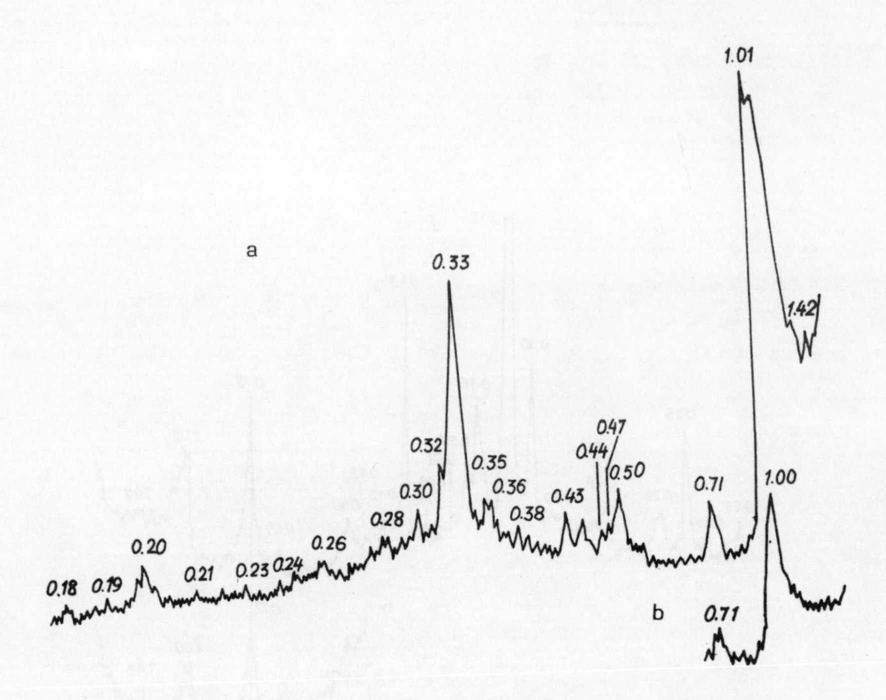

Fig. 26. Diffractograms of rocks of the Georgievian formation (a—original sample, b—sample soaked in glycerine). Igol' oil-field, well 11, depth 2799.1–2808.3 m.

type colour the rock brownish. Seams of kaolinite and uncoloured OM with sizes varying from 0.04 mm × 0.06 mm to 0.14 mm × 0.08 mm are seen aligned in nearly parallel layers. The relatively coarsely dispersed clayey matter is mainly hydromicaceous. Individual grains of fine granular silty mixture are unevenly scattered in the rock. The reflections of clay minerals are somewhat less than in the Georgievian formation but more than in the oil-bearing Bazhenovian. An admixture of montmorillonite is noticed as a result of transformation of volcanic ash, present in the form of a very thin constituent in the rocks. The microtextures are nearly axial (Fig. 28a).

In the diffractogram of a sample from a depth of 2780.5–2791.5 m, the first basal reflection of the hydromica is not high and the form of the curve does not change even in the sample treated with glycerine. The microtextures exhibit a blocky character and within the blocks there is no orientation (Fig. 28b). In this sample a new formation of kaolinite along the hydromica inside the body of OM of the second type is negligible.

The rocks of the Bazhenovian formation from the Chertalin structure resemble in type those of the Tutleim north of Tyumen district where they screen resources of gas. We think these rocks could be good screens of oil-pools in the pre-

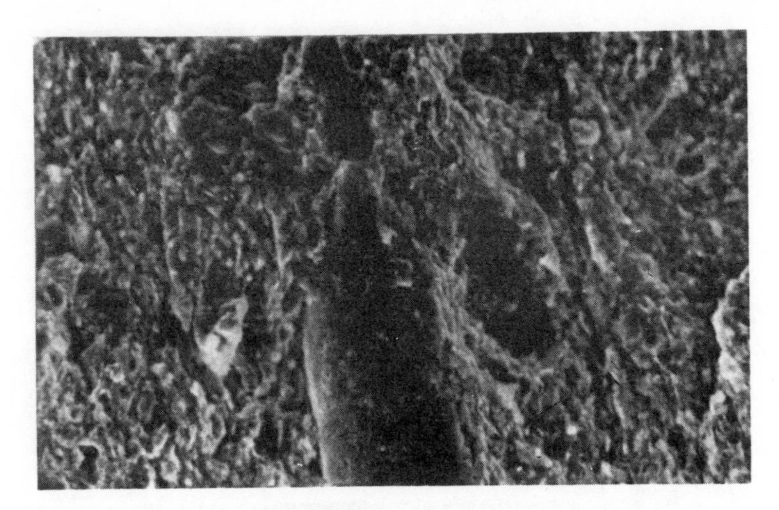

Fig. 27. Microtexture of rocks of the Bazhenovian formation. Igol' oil-field, well 11, depth 2792.1–2799.1 m; magn. 300.

Jurassic rocks if they were opened up.

The Sel'veikin structure is situated in the northern extremity. The Pudin mega-arch lies parallel to the Eastern Moiseev structure. The Bazhenovian formation contains hydromica, highly dispersed clay, and OM primarily of the first and second types generally making up 7.55%. The mesotexture is indistinct lenticular. A new formation of kaolinite after the hydromica is seen inside the body of OM of the second type. Individual crystals of calcite are also noticed (Fig. 29, I). OM of the first type is almost completely pyritised. Pyrite and crystals of calcite form centres of stresses around which the enclosing microtextures have formed (Fig. 29, II). These rocks cannot serve as reservoirs. They can only serve as the cap rocks or covers for oil and gas accumulations in the intervening complex.

An analysis of the materials from the clayey reservoir rocks of the Bazhenovian formation of the Salym oil-field has helped in formulating their characteristics:

1. The rocks of the Bazhenovian formation containing industrial oil possess a specific material composition by which they can be distinguished from the underlying and overlying deposits.
2. Textural heterogeneity appeared in the rocks as a result of the geochemical and structural interaction of the principal rock-forming components. At sites of linkage of parts of different textural types, weakened zones have formed–the basis for determining the reservoir capacity and permeability of these rocks. Disturbed under the action of tectonic stresses, the

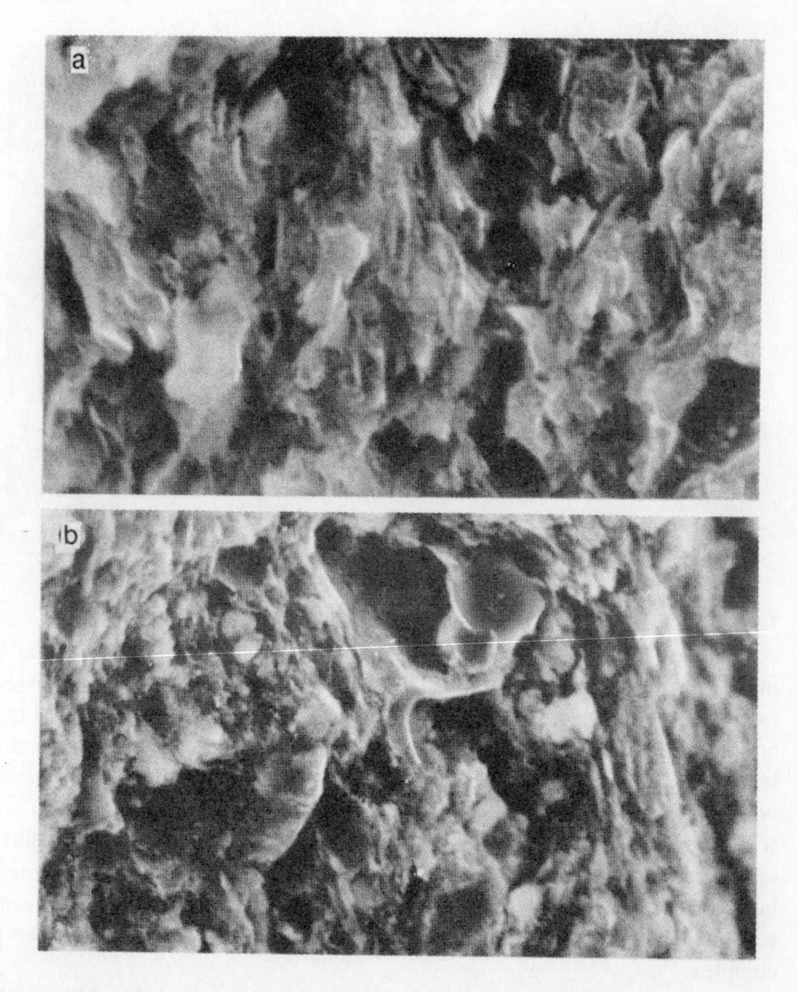

Fig. 28. Microtextures of rocks of the Bazhenovian formation of the Chertalin oil-field (well 1)
a—depth 2769–2780.5 m; magn. 3000; b—depth 2780–2791.5 m; magn. 1000.

weakened zones served as channels of migration of oil.

3. The tectonic structure of the Bazhenovian formation is determined by the block structure of the basement. The blocks are of different sizes, forms and trends, which influenced the growth and orientation of the system of faults [41].
4. Hydrophobisation of the surface of single crystals of clay minerals, and that means of all zones of contact with one another and with other

I

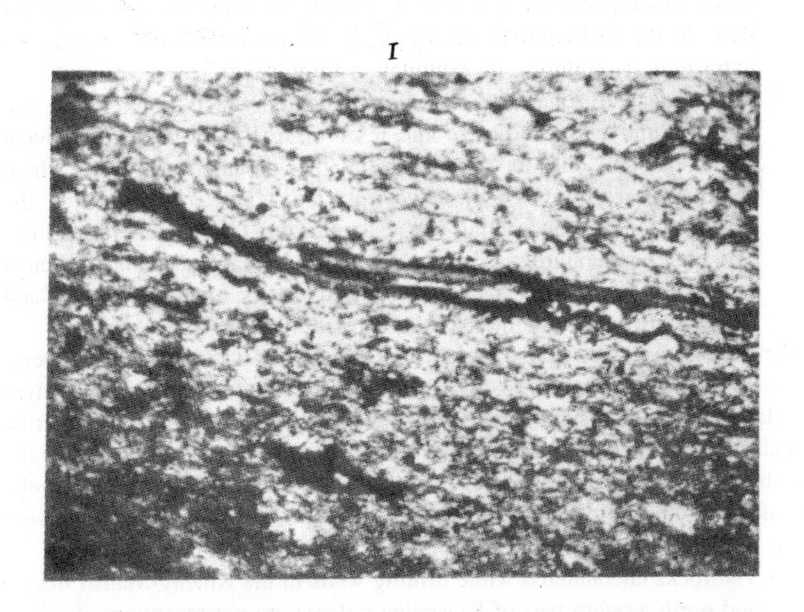

II

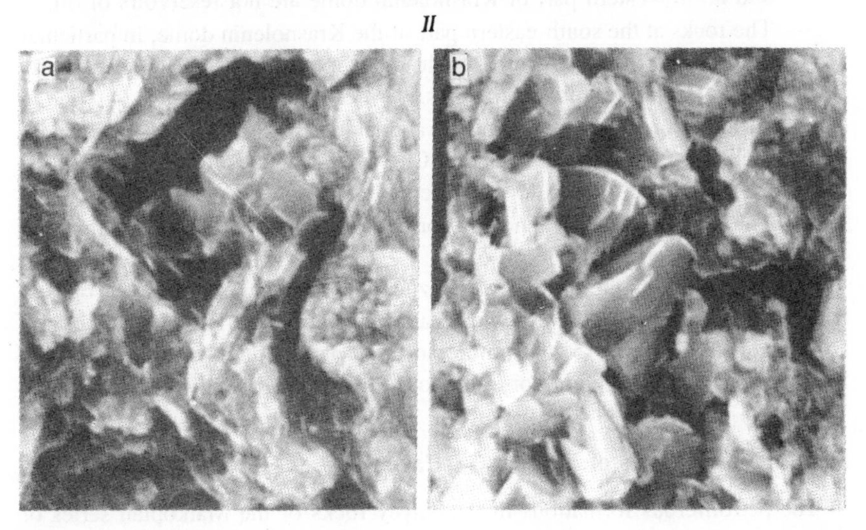

Fig. 29. Mesotexture (I, magn. 15, Nichols II) and microtexture (II, magn. 3000) of rocks of the Bazhenovian formation. Sel'veikin field, well 3, depth 2535.9–2541.1 m.
a—orientation of clay particles around pyrite concretion; b—new formations of quartz –centres of stresses around which the clay particles are oriented.

microcomponents of rocks such as the adsorbed OM, took place in the rocks. Hydrophobisation promoted easier disintegration of the constituent parts of the rocks during the filling of the reservoir with oil and also extraction of oil during its exploitation from clayey reservoirs.

5. The weakened zones are tectonic in nature, which implies that the capacity and permeability of the reservoir rocks of the Bazhenovian formation are due to the faults in the basement through which the hot gases moved [39], involving the high temperature of the Bazhenovian and underlying (for example, the Tyumenian) formations. The high temperature was responsible for the break-up of the weakened zones. These three factors–weakened zones, fault tectonics and temperature–constitute the basis of formation of the reservoirs.

The factors discussed above (weakened zones, fault tectonics and temperature) with reference to the reservoirs of the Bazhenovian formation of the Salym oil-field are nowhere of a high order in the regions studied in the western Siberian oil and gas basin and are not uniform throughout the basin. On the basis of the data obtained through studies on rocks of the Bazhenovian formation from wells opened up in other regions of western Siberia, the following conclusions were drawn:

— The rocks encountered while drilling wells in the Khanty-Mansii trough and north-western part of Krasnolenin dome are not reservoirs of oil.

— The rocks at the south-eastern part of the Krasnolenin dome, in particular over the Pal'yanov structure, might be reservoirs of not very high capacity according to preliminary findings.

— At the Nyurol' depression, as per available data, the rocks of the Bazhenovian formation of the Igol' and Chertalin structures are not at all of the reservoir type. Quantitatively, they correspond to cap rocks.

— The rocks of the Bazhenovian formation of the Sel'veikin structure can be classed as reservoirs of oil.

All conclusions on the rocks of the Bazhenovian formation from wells opened in the Nyurol' depression require verification by further borings; we agree with F.G. Gurari and others [15] that in the northern part of the Nyurol' depression, it is necessary to drill exploratory boreholes over the Bazhenovian formation.

3.3 Maikopian Series of Eastern Cis-Caucasus

The petroliferous formations in the clayey rocks of the Maikopian series of Eastern Cis-Caucasus, namely, the rocks of the Batalpashinian and Khadumian formations, belong to the Oligocene. The presence of oil in the clayey deposits was established in 1953 in some fields (Ozeksuat, Praskov, Zhurav, Posholkin and others) with initial yields of 5 to 23 m^3/day [32]. The largest oil flow was encountered in the Zhurav oil-field situated in the Alexandrov region of

Stavropol' territory. The oil-fields are located over the uplift bearing the same name (Fig. 30).

In the tectonic plan the Zhurav oil-field lies in the western part of the Eastern Stavropol' depression at the zone of its linkage with the Stavropol' anticline. The Zhurav structure is a gently sloping uplift block 18.5 km × 10 km with slope angles of the sides varying from 0°30′ to 2°. Cretaceous, Palaeogene, Neogene and Quaternary deposits comprise the structure of the formation.

Khadumian rocks in the well sections are tentatively revealed along one of the reference horizons in the electrical logging, above which are situated thin intercalations of calcareous clays; in some fields the intercalation of compact dolomite is 0.3-0.5 m thick. The lower boundary of Khadumian formation shows replacement by calcareous clays of the Eocene with characteristic microfauna, dark brown clays containing less carbonate material and impoverished organic remains. The upper boundary, between the Khadumian and Batalpashinian formations, has not been established either palaeontologically or lithologically [24].

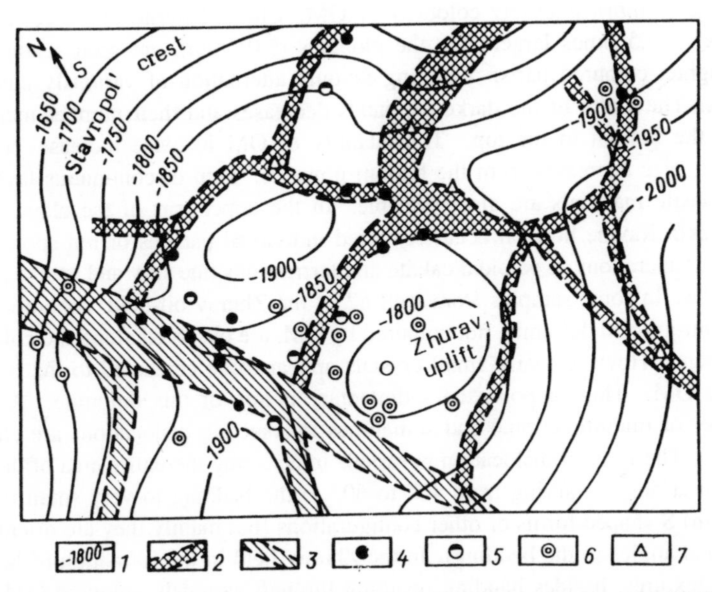

Fig. 30. Scheme of basement faults of the Zhurav oil-field (after P.S. Naryzhnyi, 1986). 1—isohypses of the roof of the Batalpashin marker, m; basement faults: 2—from data of MOGT; 3—inferred; Wells: 4, 5—corresponding to highly productive and low-yielding wells, 6—non-producing (dry), 7—recommended drivage.

The Khadumian formation in Eastern Cis-Caucasus exhibits a three-tiered division. The lower horizon–Pshekhian of the Zhurav field–consists of greenish-brown clays with lenses of clayey matter of intensely brownish colour, which are unevenly distributed but show a tendency towards layering. Many fine-grained but splendid biogenic crystals of calcite (0.01–0.04 mm) are evenly distributed in the clayey groundmass and sometimes form clusters. The content of OM is 1.35–1.85% and comprises both the first and second types. OM of the first type is thoroughly pyritised. Pyrite is present as globules, the sizes of which are comparable to the sizes of the calcite crystals. It is uniformly distributed in the rocks and in some cases does not exhibit a distinct relationship with the OM. Small planktonic foraminifers and phosphatised remains of fish are found along the bedding.

The middle horizon–the Polbinian (the ostracod layer)–is made up of brownish-grey clayey limestones in varying degrees, with the lower part dolomitised and the upper part crowded with ostracods. The thickness is 2–3 m. This is a reliable regional marker bed in the Eastern Cis-Caucasus.

The upper horizon–Morozkin Ravine–at the lower part consists of light to dark brown clays. The darker bands developed during the formation and intercalation of the major light brown clays with thin clayey layers or laminations (0.04–0.06 mm), intensely coloured by OM. Clots of pyrite of a size, for example, 2–2.5 times larger than the globules (0.01 mm), are seen scattered in the lighter-coloured parts. Layering favours alternation of variously coloured bands. The width of the darkened bands decreases and their number increases from the bottom to the top. The quantity of OM likewise increases and its particle size decreases from the bottom upwards. Thin discontinuous layers of pyroclastic materials are also traceable. In the upper part of the clays of the Morozkin Ravine horizon occur scattered individual patches or red spore fragments, concretions of peloidal calcite and terrigenous fine silty and coarse pelitic particles. In some samples from well 62 of the Zhurav oil-field, concretions of medium-grained dolomite, not coloured by OM, are observed along the coloured OM band. However, very small crystals appear to be cemented to the borders of the OM. This supports the sedimentary nature of this dolomite. A large number of multidirectional and sometimes intersecting dislocations are characteristic. The form of dislocation is varied: intersecting the entire area of the thin section at angles varying from 45° to 60° to the bedding to very small (0.4 to 0.6 mm) S-shaped forms or other configurations (but mainly they are oriented at the cited angles to the bedding). In the Zhurav field in the formation of layered micro-textures, besides banding resulting through variously coloured OM, volcanic material in the form of thin discontinuous laminae from 0.04 to 0.12 mm also contributes to and sometimes forms lenticular clots 0.4 mm × 0.6 mm. The thickness of the Khadumian formation in the Eastern Cis-Caucasus varies from 30 to 45 m.

The Khadumian formation overlaps a series similar to the dark grey clays of the Batalpashin formation, at the roof of which dolomitised limestones of 1–2 m thickness can be distinguished. This second marker bed is of Oligocene age. The thickness of the Batalpashin formation varies from 10–40 m in different regions. Microscopically, the clays of the Batalpashin formation exhibit the same colour as those of the Khadumian but are more coarsely dispersed with intercalations of medium- to fine-grained silt in the upper part. An admixture of fine-grained (coarse pelitic) silty particles is observed in the undoubtedly clayey part. Terrigenous particles are comparatively uniformly distributed. Inside the clayey part, the independent layers are darker in colour because of patches of OM of the first and second types and also pyrite concretions.

Detailed microscopic studies of the clayey reservoir of the Maikopian series above all highlight the low content of OM, which rarely exceeds 3% (data from the analytical laboratories of IG and RGI). OM is represented by all the three types but OM of the second type is predominant. OM of the third type is present in minimum amounts except in the Pshekhian horizon, where it is negligible. The proportions of OM of the third type increase from the bottom to the top along the section but in the Morozkin Ravine horizon this type does not even constitute 20% of the general C_{org} content. The OM of the second type in the deposits of the Maikopian series is characterised by a small amount of hydrolysed components, which favour its considerably larger dispersion than in other clayey reservoirs.

It is essential to bear in mind the fact that though the OM content in rocks of the lower Maikopian is not high, the percentage of chloroform *bitumenoid* in OM is fairly high, particularly in the rocks of the Khadumian formation. We present here the data from well 67 of the Zhurav oil-field for a comparative analysis. In the clay sample from the Batalpashin formation with C_{org} 2.33%, the yield of chloroform bitumenoid was 0.44%, but in the upper part of the Khadumian formation in clay with C_{org} 0.86% the bitumenoid was 0.38%. Throughout the entire Khadumian formation the C_{org} varies from 0.92 to 2.75% and the bitumenoid from 0.34 to 0.72%.

Concentrates of OM in the rocks of the lower Maikopian were also studied by the thermoanalytical method in order to determine the composition of OM and the degree of its transformation. This study was carried out by N.M. Kas'yanova. It was established that the OM in these samples exists in the lower (Khadumian formation) and middle (Batalpashin formation) stages of transformation.

The DTA curves of the concentrates of OM from the rocks of the Khadumian formation are typical of subcolloidal OM characterised by almost similar peak values at 330°C (dissociation of volatile fractions) and 460°C (dissociation of non-volatile fractions), while the latter may be complicated by 'shouldering' at a temperature of about 440°C. Compositionally, the OM consists of euxines, spores and cuticles with a preponderance of luxinite. The loss of mass during oxidation of volatile fractions was 35% and of non-volatiles, 65%. The general loss on pyrolysis was 67%. This OM provides a high capacity for generation of liquid hydrocarbons (HC).

The DTA curves of OM concentrates from the clays of the Batalpashin formation are typical of cellular-vegetal OM wherein compositionally vitrinite predominates. The DTA curves exhibit symmetrical low temperature peaks at 330°C and a high temperature peak at 460°C, with 'shouldering' towards the region of high temperature. The loss of mass during oxidation of volatile fractions was 25% and of the non-volatiles, 75%. The general loss on pyrolysis was 48%. This OM possesses a somewhat lower capacity for generating liquid HC.

The rock-forming clay minerals in the deposits of the Lower Maikopian are hydromica and kaolinite, which vary in quantity according to the section. Mixed-layered minerals of the type hydromica-montmorillonite are present as admixtures with insignificant swelling phases and chlorite. The low content of OM of the third type in the rocks was almost not revealed in the reflections of the hydromica in diffractograms. Only in the rocks from well 62 of the Zhurav oil-field, in which high yields of oil are obtained, were the reflections of the clay minerals low and some of them halo-forming (Fig. 31). In the rocks of other wells studied, the hydromica exhibited its usual character of reflections.

Unlike the clayey reservoirs of the Bazhenovian formation, the Domanikian horizon and other deposits, the clayey rocks of the Lower Maikopian do not give the characteristic maxima in the apparent resistivity curves (AR) in the range of high oil productivity. The maxima in the AR curves are related only to the carbonate intercalations. The curves of radioactivity logging are also not characteristic and hence cannot be used for correlation of sections and delineation of productive zones.

The next rock-forming component in the rocks of the Lower Maikopian is silica. Silicification of the rocks took place much faster in the second half of diagenesis, when the clayey sediments lost considerable amounts of water, such that it permitted precipitation of silica from the solution. This precipitated silica occupied only the interstices between the clay minerals where the quantity of water was still sufficient or, in other words, where there was enough space for its crystallisation. At the same time, it also formed fragments around the microblocks and microaggregates of clay minerals, reducing their hydrophilic character. The specific pattern of distribution of silica in the clayey rocks predetermined the character of dislocations appearing in the rocks during tectonic stresses, when along the boundaries of dislocations unique shiny surfaces appeared, very much resembling the slickensides that appeared during the shearing of siliceous metamorphic schists.

Silicification played a significant role in the formation of the useful reservoir capacity of the clayey rocks of the Lower Maikop. The silica in them substituted for OM in giving rise to the textural heterogeneity of the rocks at meso- and microlevels and also to the unique hydrophobised film over the clayey aggregates. The character of interaction of silica with the different components of the clays is

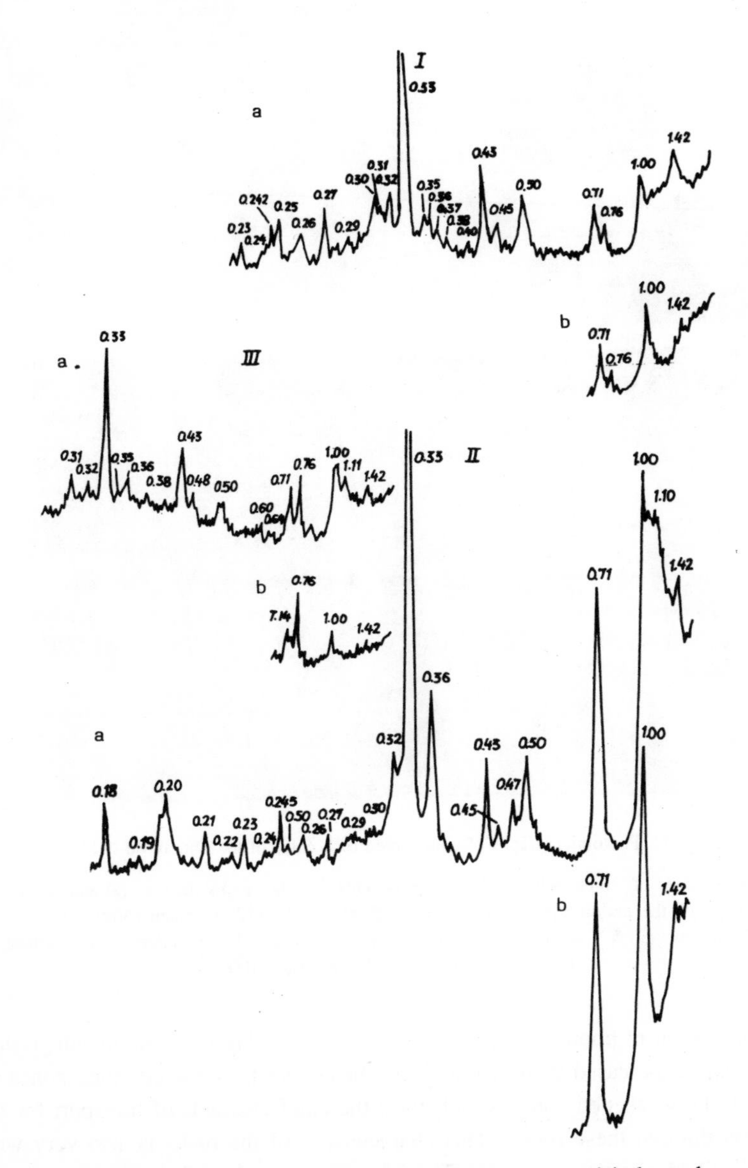

Fig. 31. Diffractograms of rocks of the Khadumian formation (a—original sample, b—sample treated with glycerine). Zhurav oil-field, well 62, depth 2112–2120 m, (I); 2152–2156 m (II); 2104–2112 m (III).

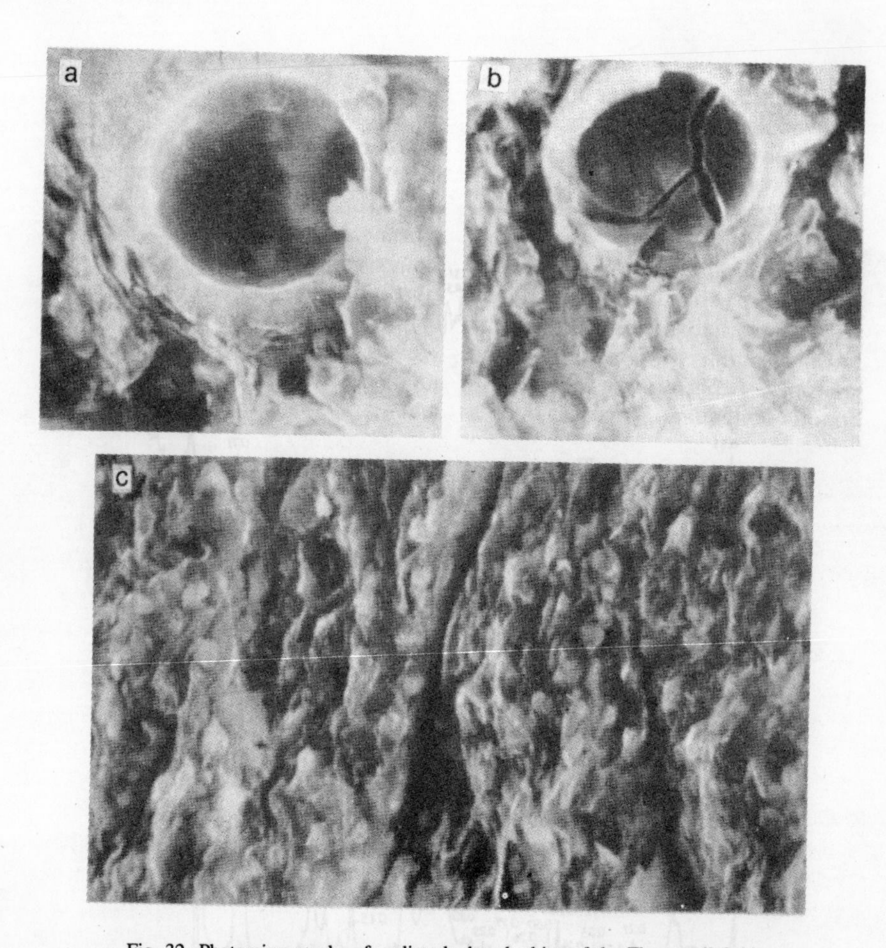

Fig. 32. Photomicrographs of undisturbed rock chips of the Zhurav oil-field.

a, b—interaction of silica with various components of the rocks before (a) and after (b) dehydration of the collomorphic OM (well 64, depth 2104–2112 m; magn 1500); c—globules of collomorphic silica formed during late diagenesis which did not undergo recrystallisation (well 62, depth 2152–2165 m; magn. 1000).

distinctly seen in photomicrographs (Fig. 32). The clay minerals disintegrating due to silica increased their characteristic behaviour to break up along zones of contact–the weakened zones–which form the chief channels of transport for oil and gas through these rocks. This characteristic of the rocks is also very well seen in SEM photomicrographs (Fig. 33). The particles of clay minerals with adsorbed silica poorly disintegrate, thus preserving the high aggregate character of the rock. Additional reservoir capacity was provided by the orientation of clay minerals around the centres of stresses, which are represented by the small amounts of quartz grains, calcite and dolomite crystals present in the rocks and also the contacts of microblocks of hydromica and kaolinite.

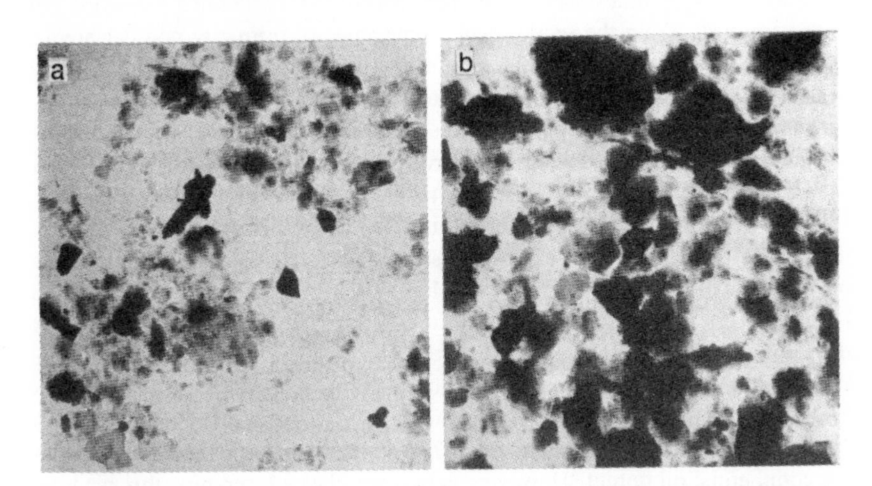

Fig. 33. Photomicrographs of fractions < 1 μm of rocks of the Lower Maikopian from the Zhurav field, well 62, depths 2178–2190 m (a) and 2145–2152 m (b); magn. 5000.

The capacity of silica to replace OM in the hydrophobisation of the surface of the clayey aggregates helped silicon to form complexes with OM and various metals during high temperature and pressure. The complexes of silicon with OM gave considerable stability because of the presence of independent $3d$-orbitals in which the small amounts of adsorbed OM and metals were firmly fixed, preventing their escape. This characteristic of silica was well brought out in experiments by B.I. Maevskii in 1983.

It is evident that the silicification of the rocks of the Lower Maikopian took place at the end phase of the diagenetic stage of transformation of the sediments, as shown by the absence of the radiolarians–the major silica builders in other types of clayey reservoirs. The time of silicification of the rocks and also the presence of traces of pyroclastic material support the endogenous source of silica in the rocks of the Lower Maikopian.

The palaeogeographical situation at the time of accumulation of the petroliferous part of the Lower Maikopian rocks has been deduced as follows, based on the data of B.A. Onishchenko [29]*. In the Khadumian period there existed a marine basin bound on the north by the Cis-Caucasian dry land plains and on the south by the Trans-Caucasian hills. From the beginning of the Batalpashinian

* Reference number should read [31]—Language Editor.

period, to the north and south-west terrigenous materials started drifting from new sources in the Eastern Cis-Caucasus so as to determine not only the change of the facies, but also their spatial distribution. The calcareous clays with intercalations of clayey limestones in the upper part of the Khadumian formation got mixed upwards along the section with homogeneous, primarily non-calcareous clays of the Batalpashin formation. After the deposition of this series of clays in the Maikop basin, the sandy-silty material stopped reaching the basin.

As in all the clayey reservoirs described above, the weakened zones constitute the major factor of influence of the reservoir potential of the rocks of the Lower Maikopian. These zones are not just fixed by standard laboratory practices. Hence, laboratory determinations of the filtration and reservoir properties of these deposits do not reflect their effective reservoir potential. According to the data of V.F. Markov, G.N. Chepak, A.N. Markov and V.S. Kosarev, the exposed porosity of the clayey rocks fluctuates from 0.2 to 14.6%. Among the rocks of the Lower Maikopian, the upper part of the Khadumian and the lower part of the Batalpashin formations possess the maximum porosity, from 12 to 14%, and permeability (4.5–5.5), 10^{-3} μm^2. Such a wide range in the section is accounted for by the extreme textural heterogeneity. Therefore, precisely this interval provides the commercial oil output. However, as in other clayey reservoirs, this range of textural heterogeneity is also close to the wells situated at different levels.

The fact that textural heterogeneity decides the reservoir capacity of the rocks of the Lower Maikopian is borne out by data on the character of working of the wells, during which the larger the depression in the reservoir layer, the longer the duration of pulsation. The experiments conducted by V.F. Markov and others at well 55 of the Zhurav oil-field showed that in the flow regulators of diameters 3.5 and 6 mm the well worked uniformly and stably; pulsation began when regulators of larger dimensions were used. The same authors explain this closing of the fractures at the critical zone as follows. The formational pressure (29–33 MPa) exceeds the hydrostatic pressure by 5–7.5 MPa. The coefficient of anomaly changes both along the section and the area of the field (1.28–1.35). The characteristic feature of the petroliferous formation is the absence of affluent external waters. The regime of oil formation is extensive, with a participating regime of dissolved gas.

The AR and GL data do not give characteristic maxima, based on which it would have been possible to divide the lithologically homogeneous series of Oligocene deposits. The general radioactivity of these rocks is not high. According to the data of A.I. Osipov and L.A. Keligrekhashvil, it varies from 3.3 to 15.5 pg-equiv-Ra/g. The highest values are obtained for the upper part of the Khadumian formation (8.67–10.84 pg-equiv-Ra/g) and the lower ('submarker') part of the Batalpashin formation (6.14–15.53 pg-equiv-Ra/g).

In the work of G.N. Chepak and others [32], recognition of the reservoirs was done by analysing the data from acoustic (AL) and gamma-gamma logging (GGL). The authors cited the evidence of a link between oil content and the dispersed rock particles in the section. According to their data, the reservoir formation is characterised by an anomaly of increased interval of travel time of the acoustic waves (650–700 mks/m) when its background values are 350–400 mks/m, and by density values of 2.20–2.27 g/cm^3 when its background value is 2.54 g/cm^3. The parts of increased porosity were established through AL data and the scattered ranges through GGL data. The zones of mixing of these anomalies were interpreted as indicators of the presence of a reservoir (Fig. 34).

According to V.S. Kosarev [23]*, the zone of industrial oil formation is characterised by increased tectonic activity and partly frequent change in the character of tectonic movements, which are expressed in the constant change of thicknesses of individual formations. The zones of development of reservoirs in the clayey rock series are confined to the flexural folds of rocks and highly crushed zones. This author also holds that the zone of increased fracturing constituted the framework of the Zhurav uplifted block with a wide belt (10–15 km) towards the west and north.

The oil-producing wells from the Zhurav oil-field are situated in its far west plunge in the zone of its interlink (through the trough) with the eastern slope of the Stavropol' anticline. In the wells driven over the crestal part of the fold of the Zhurav uplift there are indications of occurrence of oil [32].

The sharp change in the hypsometric position of the petroliferous intercalations is similar to the confinement of the chief occurrence of oil at the zone of junction of structures of various characters. This supports the strong link between the commercial oil output in the Zhurav field and the most intensive faults that form the marginal parts of the Earth's crust. The eastern slop of the Stavropol' anticline, with which the Zhurav structure is connected through the western part of the Eastern Stavropol' depression, is broken up by a series of faults of various ages and amplitudes. The border region of the Eastern Stavropol' depression is also complicated by numerous faults which, in the overlying deposits, mainly form horst-like upthrusts, which require detailed studies to ascertain the presence of oil.

The linking of the oil formation of the Zhurav structure with fault tectonics is further supported by rampant silicification of the rocks of the Khadumian and Batalpashin formations, particularly of the parts which contain oil. Silicification was the basic factor in the formation of the reservoir potential of these rocks because they contain only a little OM in a soluble form, which formed the hydrophobic fragments in the clay minerals. During this, the late diagenetic process in the sediments of silica led not just to hydrophobisation of the most adsorbed large capacity minerals, but also increased the temperature of

---

* Should read [24]–Language Editor.

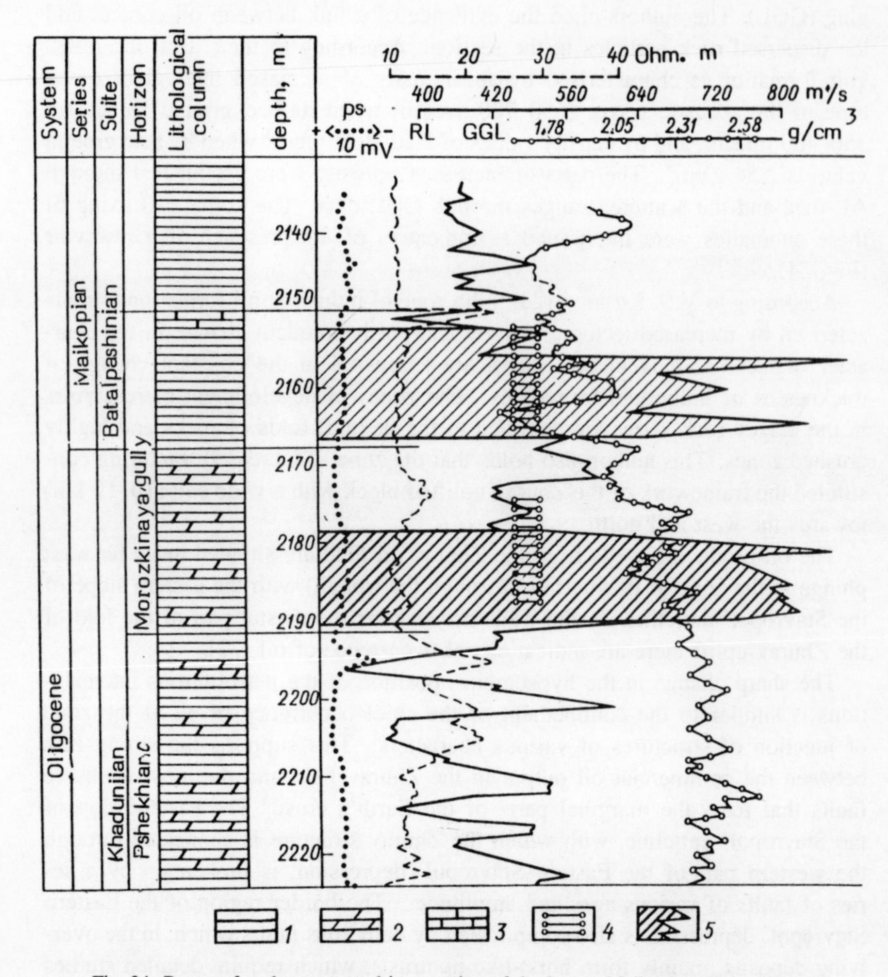

Fig. 34. Range of rocks penetrated in a section of the Lower Maikopian of the Zhurav oil-field [30].

Laminated clays: 1—non-calcareous; 2—weakly calcareous; 3—marls and limestones; 4—testing of well; 5—productive zones.

the sediments (present temperature is higher than 100°C). This facilitated the transformation of OM and also additional hydrophobisation of the clay minerals again by the hydrocarbons formed.

The author concurs in the suggestion of V.S. Kosarev that detailed structural investigations are essential for a reliable mapping of the faults that embrace the zones of intensive fracturing, to which they are related. In these parts the weakened zones at the boundaries of different textures opened up, forming the channels of migration of oil during its accumulation in the reservoir and finally in the withdrawal of oil incorporated in the reservoir.

In summing up the results of studies on the clayey reservoirs of the Khadumian and Batalpashin formations, the following can be stated:

1. Clay minerals and silica constitute the major rock-forming components of these formations.
2. The organic matter (OM) of the rocks of the Khadumian and Batalpashin formations can not be considered a rock-forming component. In the productive ranges, OM of the third type constitutes only 20% of the general content of C_{org}. OM occurs in the lower stages of transformation and has still not attained its oil-producing potential.
3. The deposits of the Lower Maikopian do not give the characteristic maxima in AR curves. Nor are the curves in the radioactive loggings characteristic. The lower natural radioactivity of the rocks of the Lower Maikopian is facilitated by the low OM content, particularly of the humus part, which is an active adsorbent of radioactive components.
4. The useful reservoir capacity of these rocks was created by the presence of weakened zones at the boundaries, wherein late diagenetic silica crystallised. An additional volume at the microlevel was formed during the hydrophobisation of the contacts of OM and silica, which served as a special substitute of OM in the rocks of the Lower Maikopian.
5. The rocks of the Lower Maikopian are characterised by the typical absence of radiolarians. This shows that silica did not reach the basin from the waters in the basin but was already present within the sediments themselves.

3.4 Menilitovian Formation of the Cis-Carpathian Foredeep

The rocks of the Menilitovian formation are sedimentaries of the Carpathian geosynclinal basin of Oligocene age, for which a high content of OM and silica are characteristic. They are widely developed and comprise the thickest and most complete sections in the northern part of the Skibov structural-facies zone, where O.S. Vyalov has distinguished three members: the lower, the middle and the upper. The thickness of the Menilitovian formation in the most complete sections is 1700–1800 m. The lower and upper members are the most enriched in OM. They are represented by dark brown to black bituminised silicified clays, with thin horizontal macro- and mesolayering. In the Menilitovian formation are found parts of layers and intercalations of tuffs,-tuffites, bentonite clays and diagenetic concretions (size 0.5–12 m) of ferruginous dolomite and, sometimes, sideroplesite. Three thick dark grey, commonly black silicified horizons containing lenticular inclusions of light grey and white silicites with an admixture of peloidal calcite comprise the formation [7].

The interesting characteristic of the Menilitovian formation is the difference in the composition of OM of the clayey and silicified intercalations. In addition to sapropelic OM, humic OM is present in considerable proportions in the clayey layers while only sapropelic OM is present in the silicified intercalations. In our opinion, this is associated with different conditions of sedimentation, as a result of which the silica of the silicites accumulated in the basin much faster than the clayey material with humic OM. There can only be one explanation for this. The accumulation of sediment with humic OM in the clayey part of the rocks took place during the mobilisation of sedimentary material from its source and hence proceeded very slowly, which facilitated its fine dispersal. Periodic vulcanicity, particularly periods of peak activity, did not last long and gradually the process attenuated. But during the short intervals of intensive vulcanicity silicified intercalations formed from the volcanic products. The material supplied by the continent was so small, however, that its presence is in no way reflected in the composition of the silicites.

A comparison of the sections of the Menilitovian formation is very difficult, particularly in such cases where the rocks of the Middle Menilitovian (Lopyanetsian) member are missing. In the case of their absence, it is difficult to divide the thick series of black clayey rocks. The difficulty is further compounded by the quick thinning of the various pockets and layers and also numerous tectonic faults. Hence A.V. Maksimov and L.M. Reifman have suggested the use of marker horizons–pockets of siliceous and calcareous rocks–for the classification and correlation of the sections.

The deposits of the Menilitovian formation are characterised by increased natural radioactivity which, as L.M. Reifman and his colleagues [36] have pointed out, is due to the enrichment of their juvenile material. Increased AR values are also characteristic of the deposits, the maximum values being confined to the Lower Menilitovian member.

Under the microscope, the clayey rocks of the Menilitovian formation exhibit fine horizontal or lenticular interbedded intercalations of various degrees of siltstones containing microfauna and glauconite.

The lower part of the Lower Menilitovian member in well 17 of the Skhodnitsa oil-field begins with a weakly silted clay layer in which, for example, layers of varying thickness of dark reddish-brown clays are sandwiched. The clays contain glauconite grains, OM of the second type and an admixture of silty particles and light brown clays, with individual inclusions of much smaller silty particles. The glauconite grain compares well with the silty grains in size. The silty admixture is more in layers with glauconite than in layers without it. The silty admixture is distributed uniformly in the clay mass, forming no segregation that gives rise to a lens or layer; the C_{org} content is 4.1% and silica 51.4%.

In the rocks of the upper part of the member, layers (0.7 mm) of finely dispersed clays with a little admixture of silt grains were replaced by clays without an admixture of silt (0.3–0.9 mm) and later by highly silted (1.8 mm) clays, changing over at places in the strongly clayey siltstone (aleurolite) again to clays (4 mm) with small quantities of silt impurities (3–5%). Filamentous laminae (0.2 mm) of finely dispersed clays, intensely coloured by OM, spread through the siltstone. In other parts, the strongly silted clays, intensely coloured by OM, changed over along the borders to siltstone or aleurolite (0.2 mm) with almost no admixture of clays, and thereby possess an almost quartzite-like structure. The layers described also vary according to coloration, indicating differences in the amount of OM adsorbed. In general, the C_{org} varies from 10.5 to 3.5% [*sic*] and silica from 31.4 to 60.6%. Usually, large layers of clayey material without an admixture or with small amounts of silty material are lenticular in form, thus giving the rock a lenticular-layered mesotexture. Layers of strongly silted clays in different parts of the rock are reddish-brown. Glauconite is concentrated in the dark brown parts of the strongly silted clays and particularly in the silty laminae. Glauconite is similar in size to the grains of quartz and feldspar, which support its allocthonous origin. In shape, the glauconite grain is rounded but sometimes has defects.

The rocks of the Lower Menilitovian member differ in OM and silica contents. The following pattern is commonly observed: the less the C_{org}, the more the silica; the more the silty laminae, the more coarse-grained the laminae; there are no layers of finely dispersed clays with a lentoid texture. Large grains of volcanic glass are encountered in various degrees of transformation, which along with the silty particles serve as the centres of stresses for the formation of enveloping microtextures (Fig. 35).

In the Ulichno oil-field at well 24 the rocks of the Lower Menilitovian member comprise dark brown clays within which lenses of light-coloured clays are distributed. The C_{org} content is 9.12% and silica 44.2%. The rocks are crowded with organic remains, radiolarians and small foraminifers. They consist of two parts: one intensely coloured OM part that contains large organic fragments and the other less intensely coloured part containing smaller fragments. Individual grains of glauconite are also observed. There is almost no admixture. One grain of glauconite was found to be filled in the centre with pyrite although its size and form undoubtedly support its allochthonous origin.

In the oil-field Volya Blazhev at well 17 the rocks of the Upper Menilitovian member consist of strongly silted clay (terrigenous admixture, from fine sand to fine silt), parts grading to clayey siltstone containing plentiful glauconite, small individual crystals of muscovite mica, remains of single- or double-chambered foraminifers and silica with a thin calcareous border 0.02 mm wide; C_{org} content 49% and silica 51%.

In the rocks of the Upper and Lower Menilitovian members, silica occurs mainly as very small lenticular concretions, with which the entire clay mass is impregnated. Part of the silica is recrystallised to chalcedony that some-

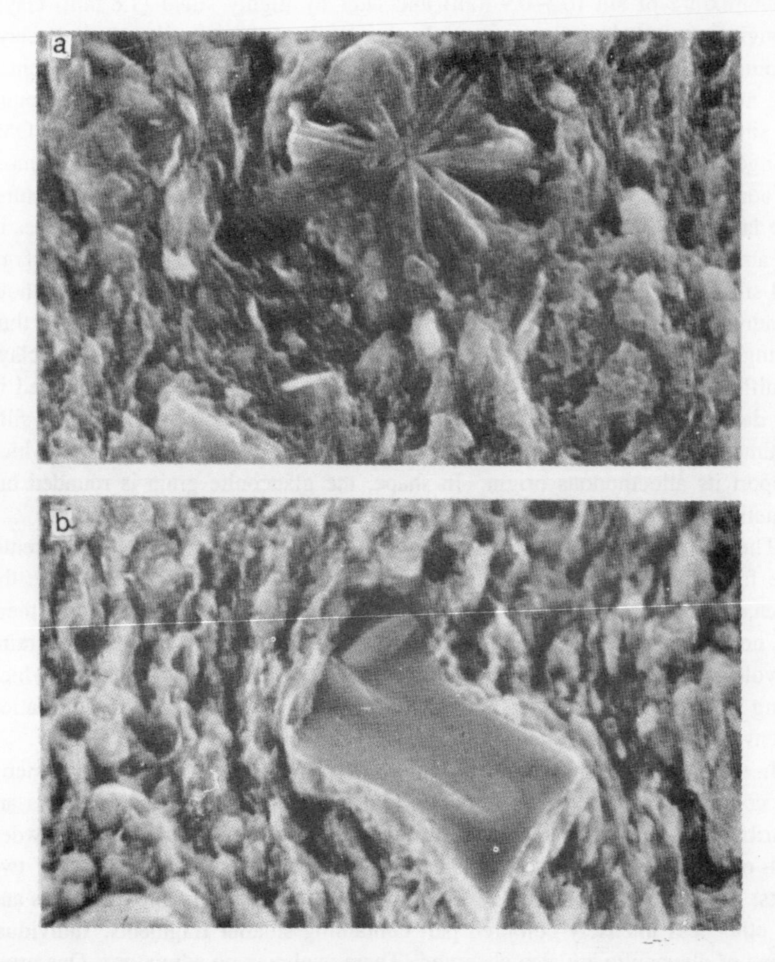

Fig. 35. Fan-shaped concretion of chalcedony (a) and fragment of volcanic glass (b) serving as centres of stresses.

a—Ulichno oil-field, well 24, depth 3446–3454 m; magn. 1500; b—Skhodnitsa oil-field, well 17, depth 4912–4920 m; magn. 1000.

times forms fan-shaped aggregates (see Fig. 35a) or occurs as siliceous pockets amidst the clay minerals (Fig. 36). Silicification of the rocks of the Menilitovian formation did not take place at just one time, as in other types of clayey reservoirs. In the first stage, during which considerable flooding of the sediments occurred, firstly the shells of foraminifers and other planktonic organisms were entrapped and their form preserved during the consolidation of the sediments. Silicification of the shells started from the outside and later along the available

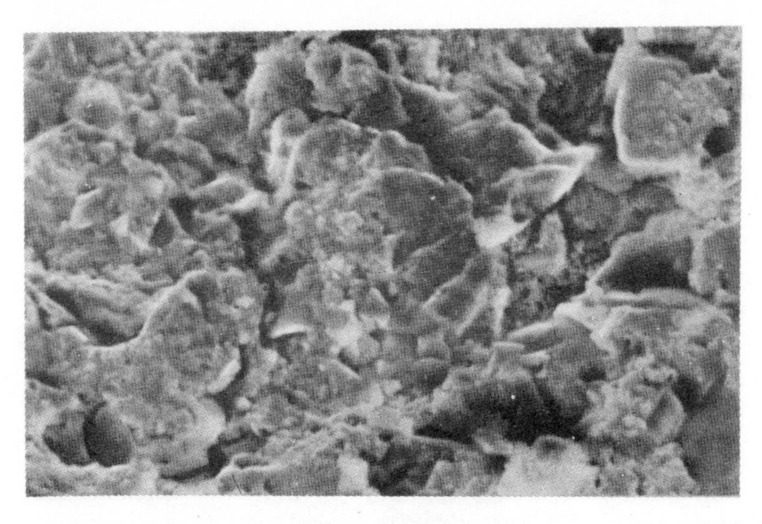

Fig. 36. Siliceous pockets in microblocks of clay minerals. Skhodnitsa oil-field, well 17, depth 4912–4920 m; magn. 480.

defects in the initial silicified zone, silica reached the inner parts of the shells (Fig. 37). At the same time, pockets of silica formed within the clay minerals. In the much later stage of diagenesis, when the rocks were almost lithified, silica took its place where some quantity of water was left behind. At these places even today the finest globules of silica are seen, which have not recrystallised (Fig. 38).

All three types of OM are present in the rocks of the Menilitovian formation but none is evenly distributed along the section. Pyrite and sometimes siderite are seen, which developed along with OM of the first type. This type is present in all the rocks of the section. Also, carbon-fixing plant remains are seen, which, according to the data of M.P. Gabinet [7], were trapped at the stage of coke formation and the remnant OM at a much lower stage. OM of the second type played a significant role in the formation of the mesotextures while, at the same time, there were not very many hydrolysable components, which contribute primarily to the development of pyrite. The latter is abundant in the Lower Menilitovian member. OM of the third type constitutes the major part of the organic matter but was adsorbed unevenly by the clay minerals, with the most intensive adsorption in the lower part of the Menilitovian formation. Based on coal petrographic analysis, the OM is found to be humic-sapropelic and sapropelic-humic [17]. The degree of bituminisation of OM is high and the katagenetic change is low. This characteristic of the OM of the Menilitovian rocks is typical of all the clayey reservoirs.

The clayey part of the rocks is represented by hydromica and mixed-layered

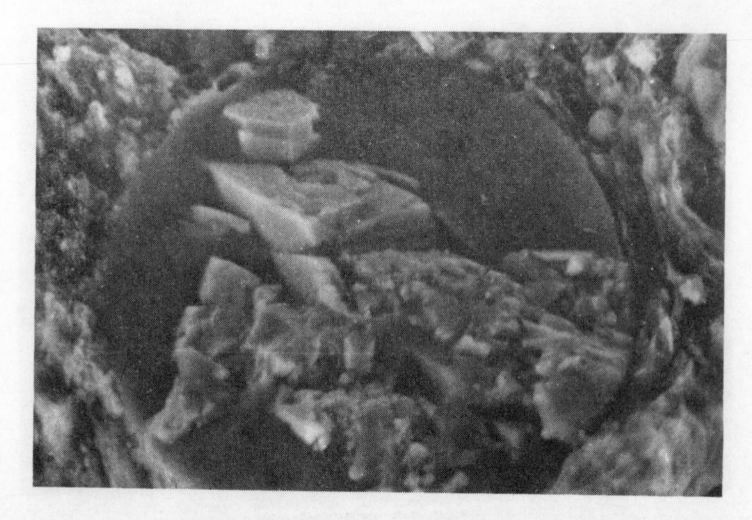

Fig. 37. Silicified shell with defects in the broken fragments from which the silicification process started. Ulichno oil-field, well 24, depth 3446–3454 m; magn. 1000.

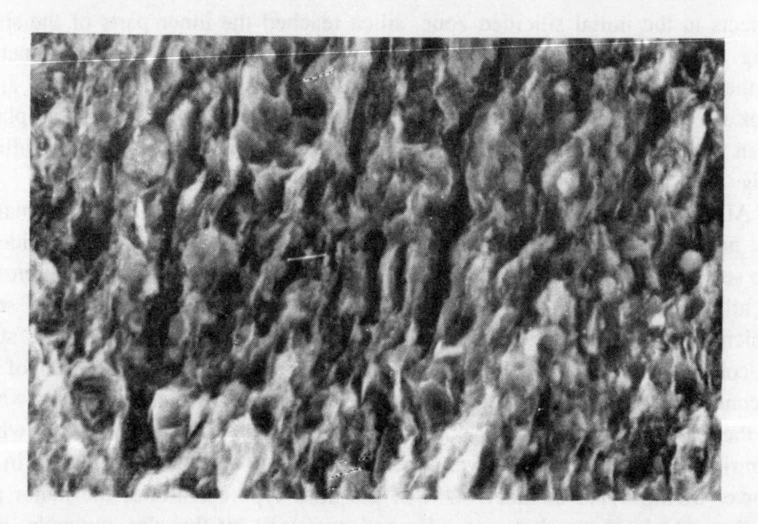

Fig. 38. Non-recrystallised globules of silica (completely precipitated in the dehydrated sediment). Skhodnitsa oil-field, well 17, depth 4912–4920 m; magn. 1000.

minerals of the hydromica type–montmorillonite. At the places of accumulation of OM of the second type, carbon dioxide was released during its transformation. In the Lower Menilitovian member of the Skhodnitsa oil-field, an admixture of kaolinite formed after the hydromica (Fig. 39, I) and in the rocks from well 17 of the Volya Blashev oil-field montmorillonite appears as an alteration product of the ash material (Fig. 39, II).

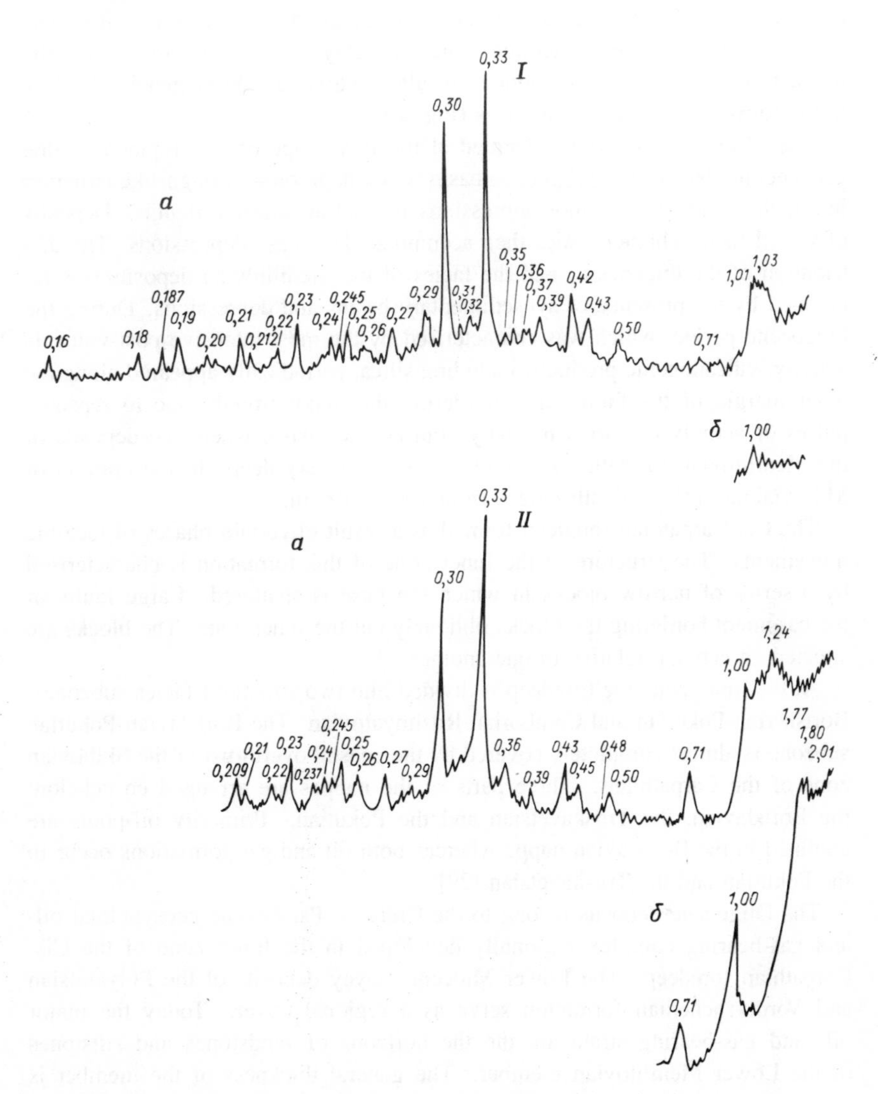

Fig. 39. Diffractograms of rocks of the Menilitovian formation (a—original sample; b—sample soaked in glycerine).

I–Skhodnitsa oil-field, well 17, depth 4912–4920 m;
II–Volya Blashev oil-field, well 17, depth 2450–2453 m.

Carbonate minerals are scattered in the rocks in the form of individual crystals. They are calcite and dolomite and rarely siderite.

Organic matter of the second type and silty layers and lenses contributed to the formation of the mesotextures, and silica and OM of the third type to the microtextures. Glauconite, ash particles, large grains of quartz and feldspars are the centres of stresses around which the clay minerals are oriented. The interaction of the mineral components resulted in textural inhomogeneity, leading to the formation of weakened zones (Fig. 40).

The Menilitovian deposits formed at the final stage of development of the geosynclinal basin. The Oligocene basin was a huge broad trough-like structure in which a number of minor depressions formed at different depths. Deposits of varied facies characteristics then accumulated in these depressions. The distribution of the thicknesses and the facies of the Menilitovian deposits was facilitated by the presence of synsedimentary basins and depressions. During the Oligocene period, which was characterised by the most intensive post-volcanic activity, vulcanogenic products, including silica, periodically appeared along the inner margin of the fault zones bordering the major trough, due to repeated pulses of activity of varied intensity. Further, the basin where the deposits of the Menilitovian formation accumulated was not very deep. In the opinion of M.P. Gabinet [7], its depth ranges from 100 to 400 m.

The Cis-Carpathian foredeep formed as a result of certain phases of tectonic movements. The structure of the Inner zone of this formation is characterised by a series of narrow blocks in which the base is shattered. Large faults of the basement bordering the blocks obliquely cut the Inner zone. The blocks are situated en echelon relative to one another.

At the Inner zone the foredeep is divided into two structural-facies subzones: Borislavian-Pokutian and Camborian-Rozhnyatovian. The Borislavian-Pokutian subzone is almost completely covered by the coastal overthrust of the Skibovian zone of the Carpathians. Three parts of the nappes are arranged en echelon: the Borislavian, the Truskavetsian and the Pokutian. Primarily oil-pools are confined in the Borislavian nappe whereas both oil and gas formations occur in the Pokutian and the Truskavetsian [29].

The Oligocene deposits belong to the Cretaceo-Palaeogene geosynclinal oil- and gas-bearing complex regionally developed in the Inner zone of the Cis-Carpathian foredeep. The Lower Miocene clayey deposits of the Polyanitsian and Vorotyshchenian formation serve as a regional cover. Today the major oil- and gas-bearing strata are the the horizons of sandstones and siltstones of the Lower Menilitovian member. The general thickness of the member is 250–400 m towards the north-west and south-east, decreasing to 60–100 m and finally, at the zone of attenuation, reducing to 20–40 m.

The Oligocene oil- and gas-bearing strata are characterised by the freshness of the sandy reservoir rocks, including those belonging to the Lower Menilito-

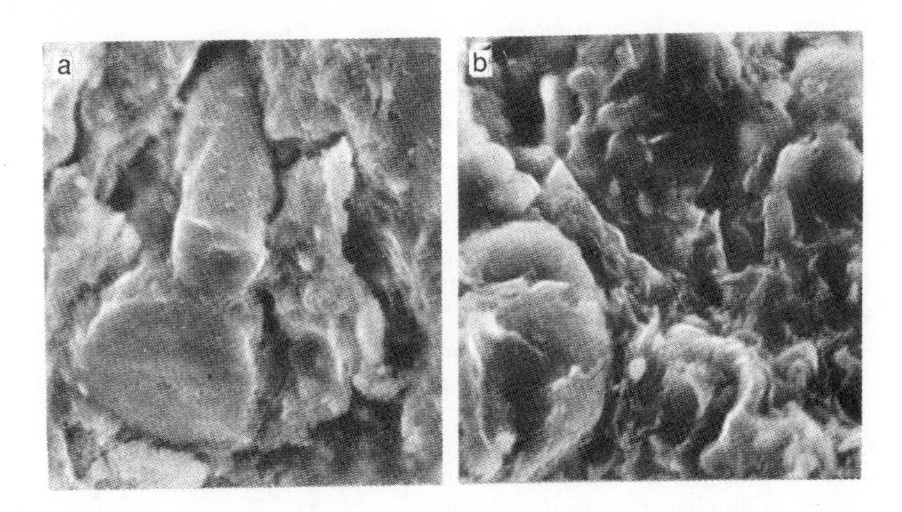

Fig. 40. Weakened zones at the microlevel in rocks of the Menilitovian formation.

a—Ulichno oil-field, well 24, depth 3446–3454 m (Lower Menilitovian member), magn. 1000; b—Volya Blazhev oil-field, well 17, depth 2450–2453 m (Upper Menilitovian member), magn. 1500.

vian member. The clayey rocks of the Lower Menilitovian member are fairly consistent both with regards to lithology and thickness. Furthermore, the specific relation between silicified organisms, silica, bituminisation, quantity of volcanic material and the depth of the basin has been established.

The capacity and filtration parameters of the clayey reservoirs of the Menilitovian formation determined on the core samples corroborate those obtained for other clayey reservoirs. Porosity is around 5% and permeability less than 10^{-3} mkm^2. It is but natural that for these rocks it is the dimension and the extension of the weakened zones and not just the porosity and permeability which decide their reservoir potential.

Primarily the clayey composition of the Lower Menilitovian subformation with OM and silica in rock forming proportions gives rise to textural heterogeneity in which the textures at the micro level contribute to the basic capacity of these rocks. Meso textures only play a secondary role. The mechanism of formation of textural heterogeneity and that means, of the weakened zones is similar to the one described for the rocks of the Bazhenovian formation. The parameters of the pore space were found to be very close. The average size of the chord of the pores varies from 0.56–2.4 mkm. The ratio of the length of the weakened zones to the unit area is found to be 1.95–2.21 mm/mm^2. The coefficient of orientation is 1.325. Transverse uplifts and depressions are found to have developed in the basement of the Outer and Inner zones of the Cis-Carpathian foredeep according to the geophysical and deep drilling data. The former are distributed in the Carpathian region through the Inner and Outer zones at the platform and the latter are situated only in the Inner and Outer zones of the foredeep [29].

Flysch sediments were deposited over the folded Riphean-Palaeozoic complex of rocks (north-west of Borislava city) and the Jurassic rocks of the platform slope (in the central and south-eastern parts) in the territory of the recent Inner zone of the Cis-Carpathian foredeep. The uplifted blocks and depressions in which the flysch deposits accumulated are bound by deep-seated faults that developed in the pre-flysch basement and attenuate towards the surface. In the flysch complex they appear as a closely spaced system of gravity faults, gravity-trans-slip faults and rarely thrust faults. The favourable material composition, facilitated by the formation of weakened zones at the contacts of textures of different ranks, and also the wide network of disjunctive faults intersecting the territory of distribution of the deposits of the Lower Menilitovian member, are the major indicators of bright oil prospects in these deposits.

3.5 Kuonamian and Inikanian Formations (Eastern Siberia)

The clayey-siliceous-calcareous deposits of the Kuonamian formation (end of Early-Middle to Middle Cambrian) are exposed in an erosional section north-east of the Siberian platform. The most complete sections are known in the estuarine parts of the Nekekit and Boroluolakh Rivers (left tributaries of the Olen'ok River) along the Greater and Lesser Kuonamka Rivers, along the Olen'ok River in a region of similar settlement, and along the Mune, Kyuyulenke and Molodo Rivers (left tributaries of the Lena River). These marine deposits of the epicontinental basin were formed under conditions similar to those of the deposits of the Domanikian horizon of the Volga-Ural region. Just in the southern part of the Kuonam basin the area of distribution of these deposits is estimated to be 650–750 thousand km^2. The thickness of the rock formation in the Kuonamian formation is 48–52 m [15, 18].

Amongst the rocks of the Kuonamian formation, the following types have been distinguished, based on material composition, as done earlier for the Domanikian formation of the Ural-Povolzh'ya: limestones, argillaceous to varying degrees, clayey dolomites, siliceous limestones and siliceous or calcareous clays, which the workers on this deposit describe as argillites. In the lower part of the Kuonamian suite, intercalations of combustible shales are encountered. Bituminisation and silicification are rampant.

Silica is present in both concentrated and dispersed forms in the rocks of the Kuonamian formation. The concentrated form of silicification is in the form of lenticular bodies varying in size from 7–10 cm up to a few metres and usually occurs as chalcedony (and rarely as quartz). These bodies are commonly distributed within the clayey-carbonate and also siliceous rocks. Dispersed silica is widely developed in the argillites, marls, limestones and dolomites. Often it appears in a cryptic form and can be determined only on the basis of the ratio of silica and alumina computed from data on the chemical silicate analysis. According to such data (based on the results of analysis of 106 samples not showing clear indications of şilicification), it was established that in 30% of the general quantity of the rocks studied, the ratio of silica to alumina is less than 5. In 70% of the rocks making up the Kuonamian formation, silica remains dispersed [15].

Studying the problem of origin of silica in the rocks of the Kuonamian formation, V.M. Evtushenko (1978) analysed the preservation of the cranidia of trilobites encountered in the lower part of the Kuonamian formation in rocks of varied composition, say from argillites and marls to limestones and silicites. Well-preserved cranidia of the same size were selected from a collection of silicified and non-silicified rocks. A spectrographic study of the chosen organisms showed that silicification had no effect on the degree of fossilisation. From this, V.M. Evtushenko concluded that the ingress of silica was penecontemporaneous with sedimentation and its redistribution in the form of lenticular bodies of silicites during diagenesis. He rightly stated that the main source of silica in the rocks of the Kuonamian formation is volcanic, primarily submarine volcanic activity. However, the time of arrival of silica in the sediments indicated by him is not wholly correct. In fact, the silica was supplied to the sediments until their consolidation but this happened not penecontemporaneously, but in the middle of diagenesis, when part of the water had already been squeezed out. Hence lenses of silica possess very odd forms, occupying places where parts were still flooded. A part of this silica formed siliceous pockets in the clay minerals and fine-grained calcite and also formed small jagged crystals occupying the pores in the carbonate part of the rock.

S.F. Bakhturov and V.S. Pereladov [2] studied in detail the sections of the Kuonamian formation in the basins of the Mune and Kyuyulenke Rivers and distinguished in them eight members: the lowermost belongs to the Aldanian stage, the overlying two to the Lenian and the five upper to the Amginian. The thickness of the Kuonamian formation in these sections varies from 32.7 to 34.4 m. These authors suggested classification of the rocks of the Kuonamian formation on the basis of the scheme of C.G. Vishnyakov and G.I. Teodorovich proposed according to the quantitative proportions of components. The rocks distinguishable in the Kuonamian formation are limestones, dolomites, argillites, siliceous rocks (silicites) and mixed rocks in which one component does not make up 50% of the rock. The limestones, for example, are classified as strongly clayey if the clayey material in them varies from 25 to 50% or clayey limestones if the clayey material content varies from 5 to 25%. On the same principle, rocks are classified as siliceous and silica-rich limestones. Argillites are subdivided into undoubtedly argillites (terrigenous admixture less than 5%), calcareous argillites (calcite 25–50%), dolomitic argillites (dolomite 25–50%) or silica-rich argillites (silica 25–50%).

The argillites are supersaturated with OM (up to 28.4%), which is seen to have impregnated the rock, sometimes forming different types of concretions and elongate laminae. The main groups of rocks in the Kuonamian formation are limestones enriched in OM and silica. The limestones are thin layered with the layers varying in width from 1 to 10–20 cm. Concretions of phosphorite 1–2 cm thick and 10–20 cm long are also encountered.

In the sections of the Kuonamian formation, the different types of rocks form a thin interbedded sequence with a preponderance in the lower part of light-coloured varieties containing low OM except for the extreme bottom where intercalations of combustible shale are exposed. The middle part is represented by combustible shales and siliceous rocks and the upper part by silicites and clayey siliceous carbonates. In spite of the wide territorial separation of the study sections, the structure of the formation appears similar in all of them [15, 18].

Within the boundary of the Udoma-Mainsk region (Yudoma, Mai, Inikan Rivers), close to the Kuonamian formation, the Inikanian formation developed, which bears structural and lithological characteristics indicative of a single age. V.M. Evtushenko [15] believed that together with the Kuonamian formation the deposits of the Inikanian comprise a unique zone of the Cambrian basin of sedimentation extending as a wide belt (up to 600 km) along the eastern part of the Siberian platform from the south and south-east regions of the recent Pri-Anabar' up to the Aldanian shield. The thickness of the Inikanian formation does not exceed 25 m.

The accumulation of the Kuonamian and Inikanian formations originated in the marine basin with normal salinity. A low rate of deposition of sediments (2–3 m for 1 million years) was characteristic of these formations. The composition of OM and the microelements associated with it changed synchronously. The quantity of microelements increased with an increase in OM content during the low rate of sediment supply. The increase in rate of deposition of terrigenous material corresponded to the decreased quantity of microelements per unit OM. It is further known that an increase in silica and carbonate content of the deposits was accompanied by a decrease in the OM content. It has also been established that the proportions of clayey and silty material decreased during the increase of carbonate content of the rocks. This pattern is but natural because the chemogenic precipitate of carbonates continued while there was attenuation in the transport of material from the continent.

In this connection, it is necessary to once again come back to the problem of the origin/source of the silica in the rocks which accumulated in the Cambrian basin. The abundant supply of silica in the Cambrian sediments is related to the Lenian period and a major part of the Amginian period. Silicification of the rocks deposited during this period is characteristic of the entire territory of distribution of rocks belonging to the Kuonamian and Inikanian formations. Analysing his projection of the possible sources of silica reaching the sediments, V.M. Evtushenko repudiated them with thorough arguments. The eastern dry land as a source of silica he ruled out because the small thickness and the immature development of the crust of weathering. The biogenic source of silica appeared unreasonable because the major silica-building organisms, diatomaceous algae, are absent in these rocks and their bloom, according to some researchers [18], started only in the Ordovician and particularly in the Silurian. In this process, significantly, the next group of organisms, the radiolarians, are encountered in the rocks in very large numbers.

The significant characteristic of the Cambrian basin is its attenuation southwards to a narrow shallow water zone of development of barrier reefs. This divides the Cambrian epicontinental basin into two parts, which differ sharply in conditions of sedimentation. The possibility of appearance and long duration of existence of such a transplatform barrier was rightly seen by V.M. Evtushenko in the presence of large ancient faults striking north-west. His point of view is supported by the presence of intercalations of plastic bentonite clays which, according to E.P. Akul'shina, are of volcanic origin. The deep-seated faults served as channels for the various volcanic products to reach the basin, providing high productivity of the plankton and enriching the sediments with silica and uranium.

A significant enrichment of uranium is characteristic of the rocks of the Kuonamian formation. A.O. Pyalling [1] carried out a comparative study of the degree of extraction of uranium from the deposits of the Bashenovian and Kuonamian formations by solutions of sulphuric acid and sodium carbonate. The proportion of uranium bound to the rock material to that of its general content in the rock constitutes 28% in the Kuonamian and 47.8% in the Bazhenovian formation. This means that the process of leaching of uranium in the acid medium from the rocks in the Kuonamian formation was more complete than from the Bazhenovian. The leaching by soda showed the amount of uranium in a mobile form: in the Kuonamian formation it was 88.5% and the Bazhenovian, 83.5%. The original uranium content in the rocks of the Bazhenovian formation was 23.5 g/t and in the rocks of the Kuonamian formation, 27.5 g/t. The uranium which migrated in solution during acidic processing (average data) from the rocks of the Bazhenovian formation was 52.5% while that from the rocks of the Kuonamian formation was 72%. The values obtained during the treatment with soda corresponded to 16.5 and 11.5% respectively.

The accumulation of uranium, silica and microelements in the rocks of the Kuonamian and Inikanian formations, in our opinion, is related to vulcanicity, which holds true for their occurrence in other clayey reservoirs. True, in the work of N.I. Matvienko and V.I. Moskvin [26] there is an indication that vulcanicity appeared in the Early Cambrian period in the framework of the basin in which rocks of the Kuonamian and Inikanian formations were deposited, which released silica, uranium and microelements. However, silica accumulated in these rocks not during sedimentation, but during diagenesis, at its middle and late stages.

Unfortunately, data from wells on the rocks of the Kuonamian and Inikanian formations and also data on their reservoir parameters estimated by conventional methods are lacking in the published literature. But the data on the extent of the weakened zones at the mesolevel, computed over the Quantimet-720 from photographs of these rocks submitted by N.I. Matvienko, support the fact that at the corresponding depths they would be reservoir rocks with satisfactory capacity parameters. In the bituminous, siliceous, clayey, carbonate rocks of the Kuonamian formation, because of the outcrops in the mid-course of the Olen'ok River (Fig. 41a), the average size of the chords of pores parallel to the bedding is about 6.4 μm and perpendicular to it 3.3 μm. The specific surface correspondingly is 0.008 and 0.016 mm^{-1}. The data reflect the small sinuousness of the pores and further reveal that in the direction parallel to the bedding it is 2 times less than in the direction perpendicular to it. The extension of the weakened zones is 284 and 543 μm and the coefficient of orientation 1.912; i.e., the number of weakened zones perpendicular to the bedding is nearly double the number of zones parallel to the bedding although, as distinctly seen in the photographs, they are clearly borne out as parallel bands.

A sample collected from an open pit in the Arga Sala River (tributary of the Olen'ok River) presents alternating thin layers of bituminous-clayey-siliceous rocks with thin intercalations of bentonite clays (Fig 41b). The average size of the chords of pores parallel to the bedding is 3.7 μm and of pores perpendicular to it, 2.7 μm. The specific surface is 0.01 and 0.012 mm^{-1} respectively, which shows that the sinuousness of the pores is greater and that their size is equal in both directions. The extent of the weakened zones is 343 and 426 μm and the coefficient of orientation 1.244.

A sample collected from an open pit in the Yudoma River (tributary of Mai River, Abdan River basin) shows bituminous-clayey-carbonate rocks (Fig. 41c). The length of the weakened zones along the bedding is 250 μm and perpendicular to it, 368 μm. The main size of the chords of pores corresponds to 3.8 and 2.9 μm. The specific surface is 0.007 and 0.010 mm^{-1}. This points out that the sinuousness of the pores perpendicular to the bedding is somewhat higher than for those parallel to the bedding. The coefficient of orientation is 1.467. This means that the number of weakened zones perpendicular to the bedding is 1.5 times greater than that parallel to it. As exemplified by the data, the weakened zones formed primarily perpendicular to the bedding. This is readily understandable because the flow of components in the sediments in the middle

a

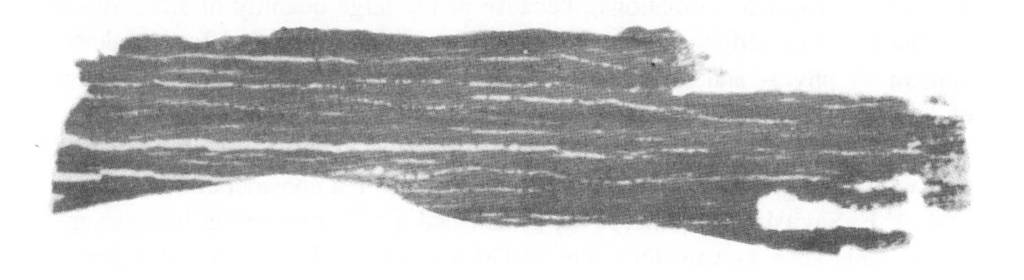

b

c

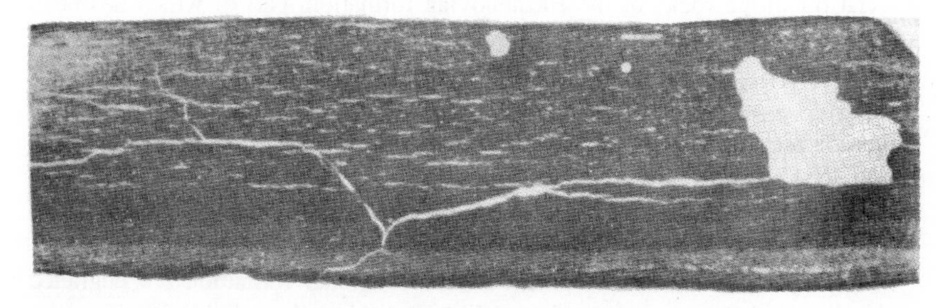

Fig. 41. Mesotextures of rocks of the Kuonamian (a and b) and Inikanian (c) formations (photographs from the collection of N.I. Matvienko).

and end of diagenesis facilitated change of orientation of their component parts in the direction of the migrating current.

In the Lenian and Ambginian period (the time of deposition of the Kuonamian and Inikanian formations), because of the large quantity of silica reaching the basin of sedimentation, favourable conditions prevailed for the development of phyto- and zooplankton and also benthic forms. A combination of low tempo of accumulation of allocthonous material from the source area and not very high intensity of carbonate formation in the basin of sedimentation with high productivity, facilitated deposition of a considerable amount of OM. V.E. Savitskii and others [18] established a high correlation between microelements such as vanadium and nickel not only with C_{org}, but also pyritic iron, which supports the common behaviour for all the clayey reservoirs established by us between the pyrite and the OM of the first type in them. According to the data of G.M. Parparova and others [17], colloalginite and sorbocolloalginite, amidst which the presence of forms resembling algae is noticed, are predominant in the Kuonamian formation. The organic matter of the Kuonamian formation and the OM of the Domanikian horizon is of the sapropelic type.

Enrichment of bitumenoids is characteristic of OM and their composition includes chloroform bitumenoid. Bituminisation of OM decreased with an increase in its concentration. A comparison undertaken by V.E. Savitskii and others [18] of the bitumenoids from the rocks of the Kuonamian and Bazhenovian formations showed fewer hydrocarbon (HC) oily fractions in the former while in the composition of the latter, the role of methane-naphthyl fractions was low but that of naphthyl aromatics high. This difference may be further assisted by a series of factors. Amongst them, in first place, is the large clay content and low OM in the rocks of the Bazhenovian formation, which facilitated a higher degree of transformation of OM due to the catalytic action of the clay minerals in this formation. Similarly, and no less significant, is the presence of commercial oil in the rocks of the Bazhenovian formation, part of which has been adsorbed by OM, which might distort the true picture of the composition of the hydrocarbons (HC) belonging undoubtedly to the OM of the rocks of this formation. A comparison of the bitumenoids from the rocks of the Kuonamian formation and the deposits of the Domanikian horizon not containing oil (data from L.A. Gulyaeva) highlights their major similarity. It is necessary to observe that the concentration of C_{org} in the rocks of the Kuonamian formation and also of the Domanikian horizon increased with an increase in clay content.

The clay minerals in the rocks of the Kuonamian formation are degraded hydromica with an admixture of mixed-layered minerals and chlorite. The carbonate minerals in the rocks of the Kuonamian and Inikanian formations are calcite and dolomite. Their ratio varies according to the section but calcite nevertheless constantly predominates.

It is mentioned in the published literature [18] that the rocks of the Kuonamian and Inikanian formations are situated at a depth of over 5 km in the central part of the Vilyui syneclise and at a depth of over 8 km in the axial part of the

Pri-Verkhoyan foredeep. Comparing these data with the results obtained on the extension of the weakened zones at the mesolevel and considering the increase in the dimensions of the weakened zones during their distension of the sizes of the pores, as in the rocks of the Domanikian horizon, it is possible to predict with accuracy the significant flows of commercial hydrocarbons from these rocks of the Vilyui syneclise. The parts which would be favourable are, in particular, those where the rocks of the Kuonamian and Inikanian formations are developed over the zones of shearing and crushing of the rocks of the basement reflected in the sedimentary cover in the form of flexures, overthrusts and other analogous structures situated at the junction of Vilyui syneclise and the border areas of the Pri-Verkhoyan foredeep intersected by faults. It is precisely in such parts that the rocks of the Kuonamian and Inikanian formations are situated at depths thoroughly accessible for boring. It is important to bear in mind, however, that before prospecting for commercial oil-pools and related types of clayey reservoirs is begun, it is necessary to undertake detailed mapping of the fault zones and to ascertain the periods of their formation. Furthermore, it should also be remembered that these rocks ought to be tested in open shafts and at all levels simultaneously, because the productive ranges might be situated at different levels. Industrial exploitation of the oil-pools in these deposits increases their prospects, which is extremely important for the future development of this region.

3.6 Pilengian Formation of Sakhalin Island

Rocks with considerable silica content are widely distributed in the sedimentary basins of the northern part of the Pacific Ocean mobile belt. Yu. K. Burlin believed that the greatest prospects for oil accumulation are related to the rocks of the terrigenous-tuffitic siliceous group of formations, which accumulated in the regions of Late Cenozoic folding and primarily at the late (mature) stages of their geosynclinal development. These groups of formations are known in other fold belts. In the Pacific Ocean belt, they are not orogenic. Sometimes they belong to the Upper Palaeogene. The terrigenous-siliceous formation is the most characteristic in this group and is represented by rhythmically built series in which are repeated siliceous clays, siliceous siltstones, gaize-like clays and their recrystallised variety with intercalations of siliceous tuffs, tuffs and terrigenous rocks with concretions of calcite and dolomite. The terrigenous-siliceous formation is encountered in different regions of Sakhalin Island and Kamchatka peninsula. These include the Kurasiian formation of Krasnogor region of western Sakhalin, the Pilengian formation of the border region of eastern Sakhalin, the Daekhurinian and Pilian formations of northern Sakhalin and the middle member of the Vetlovian formation of eastern Kamchatka.

The rocks of the Pilengian formation (Miocene) in the border region of eastern

Sakhalin are petroliferous in the Okruzhn oil-field belonging to a not-so-large anticlinal fold measuring 6.5 km × 1 km, disrupted by a gravity thrust. The thickness of the oil-bearing horizon varies from 100 to 500 m. The productive deposits covered by the clayey rocks of the Borian formation form a stratified arched trap, which is filled with oil up to the closure of the fold. According to the data of A.I. Yurochko, who conducted a detailed study covering most of the factual material relating to the characteristics of composition, physical properties and the petroliferous nature of the rocks of the Pilengian formation, these rocks are intensely fractured (fracture openings from 3 to 100 μm), which accounts for the high reservoir potential (porosity varies from 5 to 22%), filtration (average fracture permeability 0.15 μm^2) characteristic and also oil flows of up to 156 t/day.

Silica and clay minerals in different quantitative proportions assume importance as rock-forming components in the deposits of the Pilengian formation. The terrigenous admixture is indicated by plageoclase, quartz and fragments of rocks, mostly andesites. The admixture of fine silt is not much and rarely exceeds 20%. The groundmass of clay minerals is composed of mixed-layered formations of the hydromica type–montmorillonite and hydromica. The content of the latter does not exceed 10%. The quantity of clay minerals varies from 5 to 55% along the section. Silica is present in three forms: opal, low-temperature cristobalite and chalcedony. The free silica content varies from 35 to 85%.

Yu. K. Burlin considered the enrichment of the rocks with silica an 'echo' of vulcanicity in adjacent zones. This has been thoroughly verified. As already pointed out in earlier chapters, the groundmass of silica is endogenic and, in the present case, is of volcanic origin. Its arrival took place in three different stages of lithogenesis. The presence in the rocks of the Pilengian formation of siliceous tests of organisms supports the fact that silica reached at different stages not only during diagenesis within the sediments, but also in the waters of the sedimentary basin, in the form of blooming organisms of this group. Later, the organisms dissolved in the sediments and supersaturated the pore waters with silica.

It is necessary to emphasise here that the silica, which was assimilated by the organisms and later chemically precipitated, was indeed adsorbed by the clay minerals, creating a unique hydrophobic film on the surface of the microblocks of hydromica and microaggregates of mixed-layered minerals of the hydromica type–montmorillonite. In the rocks of the Pilengian formation, as in the Khadumian and Batalpashinian formations, silica facilitated strengthening of the clay minerals and their transformation in the reservoirs. This is particularly important because some amount of OM is contained in these rocks. According to the data of A.I. Yurochko, in the border wells the C_{org} content in the rock is up to 5%.

The organic matter is characterised by alinovian* composition and a high degree of bituminisation. The characteristics of the oil formations in the clayey series were delineated by Yu. K. Burlin and O.K. Bazhenova in their work published in 1985. Oil-pools are encountered in the katagenetic immature rocks with very low permeability. We only add that permeability was measured under laboratory conditions. This means that the rocks in which the reservoir capacity was favoured by weakened zones, were situated under conditions when these zones were covered.

The rocks have been classified as siliceous and siliceous-clayey, according to the ratio of siliceous and clayey material. The first type comprises those rocks in which more than 55% of the volume of the rock contains silica, whereas in the second type the silica occupies less than 55% of the volume of the rock. A.I. Yurochko chose the limiting content of silica according to the level of change of the external appearance of the rocks and their physical properties. According to the predominance of modification of silica in the siliceous rocks, gaize-like silicites are distinguished as those which contain predominantly cristobalite. Chalcedonolites are those which are composed of chalcedony with an admixture of insignificant amounts of the less stable forms of silica. Siliceous-clayey rocks contain silica in the form of cristobalite and rarely opal. In the Pilengian formation the gaize-like silicites constitute about 50%, the siliceous-clayey rocks 35–40% and the chalcedonolites 5–10%. The rocks are rhythmically interstratified. The thickness of the individual layers varies from 1 to 5 cm.

The distinguished lithotypes differ in composition and physical properties. Fracturing of the rocks was facilitated by tectonic processes, dehydration of the sediments and transformation of the less stable modifications of silica into more stable ones. The tectonic fractures in the rocks of the Pilengian formation can be separated into three distinct systems: one, parallel to the bedding and the other two intersecting the first system at 60° to 90°, forming a dihedral angle of 45°–82° amongst themselves. The thickness of the fractures in all the three systems is similar and constitutes 12–20 l/m with the opening 1–3 mm. Diagenetic fractures are oriented nearly subparallel to the bedding or the sinuous contours of the sutured joints lie at angles of 45°–70° to the bedding. The thickness of the open diagenetic fractures is 50–1050 1/m and the opening, 5–55 μm.

Electron-microscopic studies of the siliceous and siliceous-clayey rock reservoirs carried out by O.K. Bazhenova, Yu. K. Burlin, R.V. Danchenko, G.L. Chochiya and A.I. Yurochka, showed that free globules 0.8–4 μm in diameter are randomly scattered in the rock or form nodular aggregates–the globulites. An aggregate of small globules (0.1–1 μm) is also observed in the globulites and in size is close to the largest globule. The globules are spherical,

* Term not traceable; however, it seems to refer to an organic compound derived from a sapropelic source–Translator.

exhibiting different degrees of truncation. The clay minerals are distributed between the globules of silica forming globular-clotted microtextures. Increased crystallinity of the silica minerals improves the reservoir properties of the rocks, where fracture permeability sometimes reaches 185 μm^2, with an average value of 80 μm^2.

The mesotextures of the rocks of the Pilengian formation are indistinctly layered and massive. Layering is defined by the oriented disposition of the clay minerals. Organic matter of the first type, as in other clayey reservoirs, is pyritized. A comparison of the photomicrographs of the rocks of the Pilengian and Khadumian formations leads to the conclusion that silica entered these rocks at different periods of time. In the rocks of the Khadumian and Batalpashinian formations, silica arrived at the middle and end of diagenesis, while during injections silica was formed only as a chemogenic precipitate. The first portion was used up in the formation of siliceous pockets in the clay minerals and the second in the formation of dislocations at the mesolevel. There were two injections of silica in the rocks of the Pilengian formation but the first portion was the product of solution of biogenic silica and reached at the beginning of diagenesis. This difference in the timing of arrival of silica in the sediments of the Khadumian and Pilengian formations was caused by the sources (chemogenic or biogenic) of silica. It is obvious that chemogenic precipitation from marine waters was not possible due to its impoverishment in silica. N.M. Strakhov also pointed out this aspect in 1966. It has been supported by the oceanographic studies conducted by A.P. Lisitsyn, who showed that silica could only have undergone transformation in the sediments if it were a biogenic deposit. In order for deposition of organisms with siliceous tests to have taken place, there ought to have been an endogenous supply of silica precisely in the waters of the basin, as has happened during the accumulation of almost all types of clayey reservoirs except in the rocks of the Khadumian and Batalpashinian formations.

The first portion of silica in the rocks of the Pilengian formation, arriving during the solution of siliceous organic fragments, was used up in the formation of hydrophobic envelopes on the clay minerals; the second deposit, belonging to the middle of diagenesis, became incorporated in the sediments in the form of globules of different sizes. The contacts of the clay minerals with the silica coatings and also those joining the globules or globulites of silica with one another formed the weakened zones at the microlevel. These played a major role in the rocks of the Pilengian formation in creating a useful reservoir capacity. Extending at various angles to the bedding and either straight-lined or sinuous, the accumulations of silica globules are arranged parallel to one another and constitute the weakened zone at the mesolevel. Accordingly, a large part of the fractures in the rocks developed in them during the katagenetic transformation of the silica and the onset of tectonic stresses.

The strength of the siliceous coatings over the clay minerals facilitated not only the formation of the carbon-silicon bonds in the case of silica adsorption by the OM, but also the compensation of 3d-orbitals by the adsorbed elements, such as aluminium, iron, beryllium, cobalt and others, which favoured the formation of the protective inorganic film in the silica.

During electron-microscopic studies of the rocks of the Pilengian formation slit-like pores with openings reaching up to 0.5 μm were observed between the microblocks of the clay minerals. However, they revealed no significant influence on the reservoir properties of the rocks, serving simply as connections between the parts where globules had developed. Such a role is played by microfractures formed during the dehydration of minerals of silica.

Experiments carried out by A.I. Yurochko during studies on the petroliferous rocks of the Pilengian formation clearly demonstrate that the weakened zones constitute the major volume of the clayey reservoirs. Direct determinations of residual water saturation in the samples with natural saturation showed the presence of oil in the pores of the rocks, which occupied up to 62% of the volume of the pores. Submergence of the samples with natural saturation in the model of formational water led to energy displacement of oil at the expense of the counter-current of capillary water impregnating the samples. During this, an equivalent volume of oil was expelled. For a quantitative estimate of the displaced volume of oil, the homogeneous sample had to be related to natural saturation and was divided into two parts. The initial oil content was determined by the direct method in one part and the formational water in the model established in the other. After completion of the process of counter-current of capillary impregnation, which under laboratory conditions took three to four days, the residual oil saturation was determined by the same method. The coefficient of substitution was determined on the basis of the results obtained—the ratio of the substituted volume of oil to the initial. For the siliceous clays the coefficient of substitution was close to unity and for the gaize-like silicites distinctly lower. A.I. Yurochko assigned the reason for the latter phenomenon to the large size of the pores in the gaize-like silicites.

An analysis of the experimental data once again underscores the fact that pore size in a given instance is not so important. This is distinctly seen in graphs where open porosity is shown along the abscissa and the coefficient of substitution along the ordinate. In siliceous clays the coefficient of substitution is equal to unity during various open porosities, i.e., 6 and 13%. This means that the size of the open porosity and the weakened zones at the contacts of different types of textures control this process. Incidentally, this is clearly seen in a photograph of the sample in a container fulfilling the conditions of the model of formational water. The same is also true for pore size (capillary and subcapillary). If only oil saturation of the rock is determined, it [oil] could not have participated in filtration through pores of such a size.

Besides the region described above, Sakhalin-Okhotsk, western Sakhalin,

Okhotsk-Kamchatka and other basins also belong to the field of Late Cenozoic folding. The most complete sections of terrigenous-siliceous formations are known from Sakhalin and western Kamchatka. A series of elements in the geotectonic structure, the position of the formation in the section and its assignment to the inferred stage of evolution of the geosynclinal zones and also the characteristics of mesoscopic structures, led Yu. K. Burlin to necessarily relate these deposits to the single formation of the late synclinal foredeeps. The role of siliceous rocks in the formation of the geosynclinal systems is emphasised. The highest values of C_{org} are characteristic of the terrigenous-siliceous varieties of rocks. Two properties characterise the sections of the sedimentary deposits of western and northern Sakhalin and also western Kamchatka. They are the cyclic nature of the section and the silicified character of the rock. The wide development of rocks analogous to those of the Pilengian formation of the Okruzh oil-field, in the sedimentary basins of the north-western part of the Pacific Ocean mobile belt, suggests the intimately related bright oil and gas prospects of the region.

Summing up the feasible generalisations regarding the rocks of the Pilengian formation of Sakhalin Island, the following conclusions may be drawn:

1. The rocks of the Pilengian formation possess a reservoir potential analogous to the rocks of the Khadumian and Batalpashinian formations. That is, this reservoir exhibits medium filtration capacity parameters.
2. The principal rock-forming elements of the Pilengian formation are clay minerals and silica. Moreover, in the various intercalations, their quantity remains the same. The organic matter does not constitute a rock-forming component but exerts a great influence on the hydrophobisation of the surface of silica. A low stage of transformation and considerable escape of bituminous matter characterise the organic matter.
3. Unlike the rocks of the Khadumian and Batalpashinian formations, in which the main part of silica was released at late diagenesis, in the rocks of the Pilengian formation, both injections of silica were quantitatively identical but belonged to the early (biogenic deposit) and middle diagenesis (chemogenic precipitate).
4. The reservoir capacity of the rocks of the Pilengian formation includes the weakened zones at the meso- and microlevels; also, the fractures developed during the dehydration of silica.
5. Hydrophobisation of the microaggregates of mixed-layered minerals by OM and silica, and hydrophoby of the surface, which was facilitated by the formation of carbon-silicon co-ordinates, or the adsorbed metals forming on its surface an inorganic hydrophobic film, make it easy to obtain oil from the rocks of the Pilengian formation. This has been established in the experiments of A.I. Yurochko.
6. The Pilengian formation is oil-bearing at the structure disrupted by a

gravity thrust, which is characteristic of all clayey reservoirs. The setting of the fault in the structure underscores the fact that it appeared only after the structure had formed completely as a trap, which helped in the migration of hydrocarbons (HC) and the establishment of the reservoir.

3.7 Khodzhaipakian Formation of Central Asia

During exploratory prospecting activities in western Uzbekistan a not-so-thick member composed of clays, siltstones and limestones was encountered over a series of fields and was characterised by increased gamma activity. The member was called 'the member of gamma-active rocks'. This member, situated between the evaporite formations of the Gaurdakian series of the Kimmeridgian-Tithonian stage and the carbonate formations of the Kugitangtauian series of the Callovian-Oxfordian stage, was distinguished by M.E. Egamberdyev and B.S. Khikmatullaev in the Khodzhaipakian formation.

According to the data of A.M. Akramkhodzhaev and M.E. Egamberdyev, the Khodzhaipak formation overlying the eroded surface of the carbonate formation is very much analogous to the upper part of the reef structures and possesses the appearance of their sides and the top. The rocks of the Khodzhaipakian formation are dark grey to brown and are represented by peliodal, organic (algae) and organic-detrital calcareous clays, rarely with intercalations of clayey siltstones. Another characteristic of the rocks is the uniformity of alternation to a varying degree of very thin clayey, calcareous and bituminous laminae in the layers. Up to five laminae have been counted in one millimetre, thus supporting the seasonal fluctuation in the regime of sedimentation. In the carbonate clays, imprints of depressed shells of ammonites are exposed along the bedding.

The microlayered rocks of the Khodzhaipakian formation were studied by the same authors from cores of the wells of western and southern Uzbekistan. Peliodal, calcareous and clayey material is present in the form of concretions that break up the layering. The rocks are dissected by irregular fractures of different generations and varying widths which have been filled with brown peloidal matter. Sometimes spherical organic remains, resembling tests of foraminifers, are randomly dispersed in the mass. Segregations of black bituminous matter are encountered at the interbedded interstices highlighting the microlayered texture of the rocks. Concretions and individual crystals of pyrite appear, often together with OM and flattened carbonate concretions, in the rocks. Microlayered rocks with a considerable amount of clayey matter are easily split into thin soft scaly sheets. Individual blocks of rocks highly enriched in OM resemble carbonaceous shales. A part of the rocks from such members (field Gadzhak, well 14, depth range 3278–3283 m) consists of 70% organic remains and fragments of limestones cemented by carbonate material partly dolomitised and impregnated by bitumenoids. Small organic remains (complete shells and detritus) ranging in size from a millimetre to 17 mm in diameter comprise brachiopods, pelecypods, gastropods, echinoderms, segments of crinoids (?) and foraminifers. The organic detritus includes some rhombohedral dolomite (0.01–0.05 mm). Fragments of limestones (0.6–12 mm) are rounded. Loculate secretions of anhydrite are also observed. The C_{org} content varies from 0.8 to 10%.

A similar diversity of rocks is exposed in the series of dark peloidal limestones and calcareous clays in the south-eastern part of the Gaurdak hills. Here the horizon (0.5–0.6 mm) of grey organic detrital limestone is impregnated with oil and the surface of the limestone horizon is pitted and speckled with nests of Molluscan rock borers.

Indications of the presence of oil in the interbedded sutures of the bituminous limestones with high gamma activity and also of their high gas saturation are cited in the works of P.U. Akhmedov, A.G. Ibragimov and others. Such rocks occur in the fields of Urtabulak, Sardob, Zafar and Shurtan. On the GL and NGL curves, at the range of development of rock analogues of the Khodzhaipakian formation, three to seven small but prominent rhythms indicate increased gamma activity. The following pattern was observed. The values of maxima GL in the rhythms grow from the bottom to the top and in the NGL, from the top to the bottom. The authors are inclined to explain this fact as due to the reduced clay content and increased carbonate content of the rocks at the bottom of the section. The thickness of the rock analogues ranges from 0 to 60–70 m.

The deposits of the Khodzhaipakian formation and their analogues are distributed in the field as independent parts, which usually constitute the reef-like bodies. This supports the peculiarities of the tectonic history of the region where the accumulation of these rocks took place. As pointed out by A.K. Mal'tseva and N.A. Krylov in 1986, during the course of the Jurassic period an intensive structural differentiation resulted in the Turan shield. All the elements of the recent structural plan were formed towards the beginning of the Cretaceous and a large part of the discontinuous faults complicated the platform cover. Amudar'in and Northern Ustyurt were the two distinguished from the syneclise of the Jurassic. The Amudar'in Jurassic palaeosyneclise, besides the territory of the present syneclise, also embraced the more eastern region which at present forms part of the post-platform orogen–the mega anticline of the South-west Gissar and Afgano-Tadzhik basin.

A.M. Akramkhodzhaev and M.E. Egamberdyev constructed a palaeogeomorphological structure and along an isopach of 60 m differentiated three foredeeps in southern and south-western Uzbekistan corresponding to the zone connecting the south Tadzhik post-platformal orogenic field and the Turan plain: the Gadzhak in the northern part of the Surkhandarin ravine, the Khodzhaipak in the north-western part of the South-west Gissar (from Gaurdak in the south-west to Pachkamar in the north-east) and the Dzhambulak in the north. Further, along the isopach of 40 m, within the limits of the Chardzhous fault terrace, Ispanlii,

Alan and Southern Dengizkul local interreef foredeeps were observed in an area from 40 up to 200 km^2. According to the opinion of the same authors, local and large foredeeps appeared in the second half of the Oxfordian period and, changing over to lagoons, developed up to the middle of the Kimmeridgian. In these lagoons reef-like facies developed on the uplifted parts and clayey and clayey-carbonate facies enriched in OM were formed over the submerged areas.

Experience in studying the clayey reservoirs of various oil- and gas-bearing basins prompts us to conclude that the rocks of the Khodzhaipakian formation are similar to those of the Kuonamian formation in composition and conditions of formation. In its tectonic relation this was apparently a unique basin with its structure at the base inheriting both the structure of the basement and that of the underlying rocks of the carbonate formation of the Kugitangtaunian series of the Kellovian-Oxfordian. Furthermore, as rightly remarked by A.M. Akramkhodzhaev and M.E. Egamberdyev, while the relief of the bottom of the basin played a significant role in the formation of the reefs and the basinal sediments, these parts of the basin were bound by faults which served as the channels of transfer of the components, both for the reef-building organisms and for the clayey-carbonate sediments, thus favouring their high gamma activity.

The composition of the rocks of the Khodzhaipakian formation facilitated their textural heterogeneity at the meso- and microlevel. The microtextures formed during the interaction of the clayey and peloidal carbonate particles, and the mesotextures of skeletal carbonates with the silty particles and concretions of clayey and carbonate matter. A subordinate role in textural modification of these rocks was played by silica, which cemented the rock in different parts. The useful reservoir space was provided by the textural links of the various components of the rocks at the meso- and microlevel. Organic matter finds a place in two types of textures but considering its quantity (maximum 10%), one may rightly assert that its role was less significant than in the rocks of the Kuonamian formation and the Domanikian horizon. As in all other reservoirs in the clayey rock series, the rocks of the Khodzhaipakian formation may have formed as an oil- and gas-saturated layer simply as a result of migration of the hydrocarbons (HC) during the formation of the commercial oil-pool.

The presence of gas reserves in the carbonates of the Jurassic in the Kultak oil-field prompted A.G. Babaev, A.N. Simonenko and I.V. Kushnirov (1975) to divide the Khodzhaipakian formation into a separate independent productive zone. For this formation the gas reserves were computed and assigned to the C_2 category (around 10% of the general reserves of gas in the formation).

Our suggestion regarding the role of the faults separating the reef-forming zones from the depressions as the suppliers of components in these rocks, is supported by the results of investigations on the microelements by A.G. Babaev published in 1983. As is known, R. Chester thought that the ratio of barium to strontium serves as an indicator in differentiating reef deposits from non-reef deposits. The study of microelements conducted by A.G. Babaev revealed no pattern in the distribution of concentration of the indicator elements of reef formation. It was found that the ratios adduced by R. Chester as characteristic of reefs were practically encountered in both reef and non-reef horizons. Often in the rocks of the reef zone, the ratio of barium to strontium was less than unity and in the depressions, higher, which explains the large adsorption volume of the latter.

A review of the data collected enables the following observations:

— The rocks of the Khodzhaipakian formation are potential clayey reservoirs, without high reservoir indices.

— In type of formation and composition, these rocks, with some reservations, can be relegated to the Domanikian type because the silica in them is not deemed a rock-forming component.

— The formations in which, at present, oil and gas are exploited from the reef-forming deposits and which, like the Kultakian formation, are situated close to the regional faults bounding the structures of varied characteristics, ought to be analysed in terms of productivity of the rocks of the Khodzhaipakian formation.

3.8 Lower Permian Deposits of the Pri-Caspian Syneclise

The Pri-Caspian oil-gas province belongs to the coal-bearing plate of the eastern European platform. It is characterised by large depths and intensive faults of the basement reflected differently in the sedimentary cover. The major tectonic element of the plate is the Pri-Caspian regional syneclise filled with sedimentary rocks of vast thickness (20 km). The characteristic feature of the section is the thick saline series (up to 3–4 km in the primary deposit) of Lower Permian (Kungurian) age, which divides the entire section into subsaline and suprasaline structural formation complexes. The potential clayey reservoirs of the Pri-Caspian syneclise belong to the lower subsaline (Palaeozoic) structural stage within which, based on lithofacies and geochemical characteristics, three oil-gas-bearing complexes have been distinguished, of which two are terrigenous Lower Carboniferous and Lower Permian.

According to the data of B.K. Proshlyakov, T.I. Gal'yanova and Yu. G. Primenov, clayey rocks are present in all the stratigraphic subdivisions of the subsaline complex—from the Upper Devonian to and including the Lower Permian in the western part of the Pri-Caspian syneclise. These series, layers, intercalations and lenses vary in thickness (from fractions of a millimetre to 100 m) in a wide range of depths—up to 6028 m from the surface at the Aktyubin-Pri-Ural in the Binkzhal' field. Of the clayey rocks, the potential reservoirs belong to the Sakmaro-Artinian part of the cross-section of the Lower Permian.

The clayey reservoirs are dark grey to black, associated with various amounts of the silty and calcareous rocks. The mesotextures are primarily massive and the microtextures clotted and enclosing. The clayey matter of the reservoirs is coarse, scaly and represented by hydromica (35–40%) and chlorite (up to 40%). Mixed-layered minerals of hydromica-montmorillonite composition comprise up to 25% of the clayey fraction of the rock. The quantity of swelling layers does not exceed 5%. Kaolinite is rare. The terrigenous silt-sized minerals form layers and lenses varying in size from a fraction of a millimetre to a few centimetres. The carbonate content (from 5 to 30%) is represented by peloidal calcite. Sometimes splendidly facetted crystals of dolomite (0.14 mm × 0.06 mm or 0.02 mm × 0.02 mm) are seen, distributed almost uniformly in the rock. The clotted microtextures are favoured by the presence to a varying degree of changing grains of chlorite, around which large scaly clots of hydromica and the wrapping or enclosing microtextures are favoured by the orientations of the finely dispersed part of the clay minerals, mainly mixed-layered minerals, around the grains of the terrigenous components.

The C_{org} content varies from 4.5 to 5.8%, in which the amounts of sapropelic OM go up to 6.3%. OM is represented by components of the first and second types (according to the classification described in Chapter 1.3), while OM of the first type is preponderant. It consists of carbonised, finely dispersed vegetal remains in the form of isometric particles or thin (up to 0.02 mm) needles up to 0.24 mm long. The organic matter of the first type is almost completely pyritised. OM of the second type occurs as patches of spore fragments of varying size and dark reddish colour. The stage of katagenesis of MK_1 with a palaeotemperature of 110–125°C was determined at the recent formational temperature of 69–85°C by I.B. Dal'yan, T.P. Volkova, V.I. Gorshkov and A.S. Posadska in 1983 for the Upper Carboniferous-Lower Permian rocks lying at depths of 3700–4982 m. Parts of concretions of authigenic chalcedony in the rocks have sometimes recrystallised to quartz.

The subsaline clayey deposits, according to the classification of B.K. Proshlyakov, belong to the compact, strongly and very strongly compact categories. The coefficient of compactness varies from 0.80 to 0.97 (average 0.92). The relative density of the clayey rocks at a depth of about 4 km changes from 2.24 to 2.79 (depending on the content of carbonates). B.K. Proshlyakov, T.I. Gal'yanova and Yu. G. Pimenov pointed out that the degree of compaction of subsaline clayey rocks is below the computed value and is particularly sharply traced while comparing the subsaline and suprasaline rocks. The authors suggested that the main reason for this phenomenon is the presence in the upper part of the section of a thick series (at places up to 8–10 km) of rock salt of Kungurian age. The increase in thickness of the salt-bearing series is accompanied by a lessening of thickness in the subsaline clayey rock reservoirs, reduction in quantity of carbonate material in them and, consequently, their increased porosity.

To categorise the types of clayey reservoirs, studies were undertaken utilising the ultrasonic method, followed by saturating the rock samples with luminescent liquid. According to the data of Yu. K. Proshlyakov, T.I. Gal'ya-nova and Yu. G. Pimenov, there are two types of clayey rock reservoirs: the fracture type and the porous fracture type with empty space. The porous fracture type is observed in the clayey reservoirs with a considerable amount of terrigenous and authigenic mixture. Fractures are situated primarily parallel to the bedding and are either straight-linear, sinuous or en echelon. They are often grouped into zones up to 5–6 mm in width. The length of the fractures in the cores is 20 to 70 mm and the opening, from fractions of a micrometre to 2250 μm. In connection with the high density of distribution of the fractures (0.91–4.51 cm/cm^2), the clayey rock reservoirs are split by fractures which help in the determination of their gas permeability, particularly along the bedding. The fracture permeability determined over the photographs of the samples saturated with luminophor was estimated to be $(6–250)\cdot 10^{-15} m^2$. Gas permeability measured along the highly fractured variety was not higher than $3.15 \cdot 10^{-15} m^2$ along the bedding and $0.34 \cdot 10^{-15} m^2$ perpendicular to the bedding.

In the subsaline complex of the western part of the Pri-Caspian syneclise, clayey rock reservoirs are encountered in all the stratigraphic intervals. This explains the post-depositional transformation of the clayey rocks. A study of the core and borehole data of the GIS from the almost well-developed fields of Bozoba, Kenkiyak and Karatyube, enabled B.K. Proshlyakov, T.I. Gal'yanova and Yu. G. Pimenov to establish that the clayey rocks occur as an independent series varying in thickness from 15 to 200 m and are also seen as separate layers and lenses (up to 2–5 m) in the sandy siltstone rocks in the Lower Permian part of the subsaline complex. The data obtained on the lithological characteristics and deformational strength properties of the clayey rocks of the different parts of the section support the fact that the clayey reservoirs of the fractured type belong to the upper and lower parts of the clayey series of more than 50-m thickness and embrace completely the clayey rocks of less than 50-m thickness and also layers and lenses in the sandy-silty members irrespective of their thickness (Table 5).

A direct answer to the question of the commercial importance of the potential clayey reservoirs in the subsaline complex of the oil formations of Kenkiyak, Bozoba and Karatyube is not possible because of the system of probing, during which large ranges of the section of the Lower Permian (up to 74 m in well 91, Kenkiyak) were tested both as sand-siltstone and clayey formations. Commenting on the indirect indications of the commercial significance of the clayey rocks with a fractured character of reservoir capacity, B.K. Proshlyakov, T.I. Gal'yanova and Yu. G. Pimenov indicated that the films and fillings of oil along the fractures of the clayey rocks were responsible for changes in discharges during experimental exploitation of the formations, which, in general, are typical

Table 5: Correlation of proportions of carbonate part and coefficient of plasticity (K_{pl}) in the clayey rock series of the Bozoba, Karatyube and Kenkiyak oil-fields (according to the data of B.K. Proshlyakov, T.I. Gal'yanova and Yu. G. Pimenov)

Thickness m	Part of series	Carbonate proportion %	K_{pl}
< 50	Upper	$\frac{19.3 - 25.3}{22.2}$	$\frac{2.2 - 2.2}{2.1}$
	Lower	$\frac{13.7 - 25.3}{19.5}$	$\frac{2 - 2.3}{2.1}$
> 50	Upper	$\frac{11 - 19.1}{15.1}$	$\frac{1.3 - 3.2}{2.5}$
	Middle	$\frac{1 - 16.4}{9.6}$	$3 - \infty$
	Lower	$\frac{9.3 - 32.8}{18.6}$	$\frac{1.3 - 3.7}{1.8}$

Note: Numerator—limits of variations; Denominator—mean value.

of fractured reservoirs. The testing of the formations in well 93 in the Kenkiyak oil-field (depth 3935–3938, 4004–4008 and 4149 m) may serve as an example. In the flow regulator a discharge of 10 mm of oil was equal to 82.8 t/day which, in further testing, resulted in a yield not exceeding 40 t/day. An analogous fall in yields was observed in other wells too.

The elastic-deformation properties of the subsaline clay rock reservoirs determined under laboratory conditions, as shown by the aforementioned authors, support the significant tendency of the rocks towards the formation of fractures. In order to establish the directions of fracturing, the speed of the ultrasonic waves was determined in three mutually perpendicular directions and also studied by luminescent defectoscopy. A quantitative estimate of the heterogeneity of the clayey reservoirs was carried out using the coefficient of defectiveness K_d (the ratio of minimum and maximum speed of the ultrasonic waves) and coefficient of anisotropy K_a (ratio of mean speeds of the ultrasonic waves perpendicular and parallel to the bedding). The coefficient of defectiveness varied from 0.1 to 1. The closer the value to unity, the more homogeneous the rock. That is, the rock is either devoid of pore space or the pores or the fractures are uniformly distributed. The coefficient of anisotropy exemplifies the predominant orientation of the fractures. When $K_a > 1$, the fractures are vertical, if $K_a = 1$, the fractures are diagonal and when $K_a < 1$, the fractures are horizontal. Hence the coefficients indicated reveal whether the majority of fractures are situated parallel to the bedding of the rocks and whether a considerable number of diagonal fractures are present.

The potential clayey reservoirs of the Lower Permian of the Pri-Caspian syneclise do not, however, belong to any of the three differentiated types (Bazhenovian, Domanikian, Khadumian). They practically consist of two rock-forming parts: clay minerals and the peloidal carbonate material. The second component assumes a rock-forming character when its content reaches the maximum (30%). The textural aspect of the rocks at the microlevel provides an interspatial setting of microblocks of hydromica and chlorite and microaggregates of mixed-layered minerals of the hydromica type—montmorillonite. The microtextures are primarily axial and enveloping. The mesotextures are produced by the layering of terrigenous material either uniformly or lenticularly distributed in the layered varieties of clayey rocks, or randomly dispersed throughout a rock of massive texture. The carbonate matter forms nodules up to 0.2 mm in size. Weakened zones appear in these rocks at the borders where the clay minerals are in contact with concretions of carbonates and grains of terrigenous minerals. The organic matter (OM) played some role in inducing heterogeneity in the texture. Its absorbed components created weakened zones at the microlevel.

A characteristic of the clayey reservoirs of the Lower Permian is the absence of silica and the presence of carbonate material instead. The latter, as already discussed, cannot form a hydrophobic film over the clay minerals as this can only happen at the contact of the clay minerals with collomorphic silica. The bonding of clay minerals with carbonates is not strong, particularly because the loss of water molecules by them during consolidation of the sediments does not take place at one time. Hence the contacts of the carbonate nodules with the clayey microblocks and microaggregates are weakened along their entire surface. The same holds true for the contacts of clayey microblocks with the grains of terrigenous minerals. The high rate of sedimentation during the intensive sagging of the basin provided conditions resulting in the dehydration of the surfaces of the microblocks and microaggregates of clay minerals and nodules of carbonate matter under the influence of increased temperature. The separation of the weakened zones in the rocks of the Lower Permian could not be uniform throughout the entire rock series and hence these reservoirs would not possess very high reservoir indicator characteristics, in spite of the vast thickness of the potential reservoir formations.

Palaeotectonic reconstruction of the Early Palaeozoic-Middle Visean stage of development of the Pri-Caspian syneclise using regional refractive surfaces and reflective reference seismic horizons, enabled I.B. Dal'yan to establish that towards the beginning of the Early Palaeozoic-Late Devonian the Baikal basement was broken up by a system of regional deep-seated faults of sublatitudinal and submeridional trends. The fundamental structural elements–the uplifted blocks and ridges of the basement–and the meridional graben striking along the boundary of the Pri-Caspian syneclise with the Ural'sk geosyncline controlled the composition and thickness of the accumulating deposits.

The subsequent stages of tectonic development of the territory were characterised by the slowing down of the rate of subsidence because of the closing of the Ural'sk geosyncline. This led to sedimentation along the eastern border of the syneclise, of the carbonate deposits which, at the western Kenkiyak-Shotykol' carbonate escarpment, were replaced by synchronous, primarily clayey deposits 21–71 m thick, characterised by high gamma activity–up to 13.5 mkR/hr (known as gamma-activity member). The clayey analogues of the upper carbonate series are exposed in the borewells of the oil formations of Bozoba, Kenkiyak and others, where they directly rest on the rocks of the lower carbonate series.

The investigations on the genetic relation between the subsaline and suprasaline oil formations in the eastern part of the Pri-Caspian syneclise conducted by O.V. Bartashevich, I.B. Dal'yan, V.I. Ermakova, A.M. Medvedeva and V.S. Melamedova, revealed the presence here of three varieties of oil belonging to a single genetic type. The genetic unity of the oils of the subsaline and suprasaline oil- and gas-bearing complexes indicates their formation during vertical migration, which is also supported by the results of palaeophytological studies. The regional occurrence of oil and gas in the subsaline complex of rocks prompted me to give due consideration to the Palaeozoic deposits of the eastern part of the Pri-Caspian basin, which promise a high potential for prospecting of hydrocarbons.

The wide development of disjunctive tectonics and also the commercial occurrence of oil in the underlying and overlying clayey rocks of the carbonate and terrigenous deposits make me, like B.K. Proshlyakov and others, believe that the potential clayey reservoirs of the Sakmaro-Artinian part of the Lower Permian offer great prospect for finding oil in their formations. To support the proposition that the clayey rocks of the Lower Permian contain oil, it is necessary to substantiate the fact highlighted in the works of I.B. Dal'yan that the synsedimentary and post-depositional structures were formed much earlier than the commencement of migration of hydrocarbons (HC) from their places of formation in the subsaline and suprasaline deposits. As the enclosing structures and the arched uplifts experienced no conspicuous changes during the later development of the territory, undoubtedly not only the formation of reserves but also their preservation was ensured in spite of the presence of anomalously high formational pressure (AHFP). Of course, it is true that the calculations of B.M. Balyaev and A.E. Lyustikh support the young, geologically recent age of the formation with the sharp anomalous pressure of the formational fluids, which in the regional plan correspond to the regional geothermal anomalies. The 'aureoles of intrusion' of abyssal gases under AHFP envisaged by these authors in the larger plan are reminiscent of the 'aureole of intrusion' from large reserves in the thick isolated series, as proposed by K.A. Anikiev.

In the clayey reservoirs of the Lower Permian, favourable conditions to contain AHFP were created in those parts where the clay minerals were oriented around the carbonate minerals of pelitomorphic structure, which, losing their films of water, remained as partitions within which the clay minerals retained their plasticity and moisture. Such conditions appear at places of occurrence of microaggregates of mixed-layered minerals, the swelling montmorillonite layers of which only reluctantly give off the water contained in them and hence retain their plasticity and moisture content.

After analysing the available material on the potentialities of the clayey reservoirs of the Lower Permian, it is possible to arrive at the following conclusions:

— The mineralogical and structural-textural transformations of the rocks of the Lower Permian are associated mainly with consolidation, calling for changes in the rocks, at great depths, and the presence of carbonate material forming microaggregates of various dimensions.

— Mainly clay minerals participated in the formation of the texture. Those with small amounts of OM formed microtextures. Mesotextures formed via the distribution in the clayey mass of terrigenous minerals in the form of layers, lenses and concretions of calcite.

— Oil-pools in the clayey reservoirs of the Lower Permian formed during vertical migration. This has been confirmed by geochemical and palynological investigations of the oils from the subsaline and suprasaline complexes and also palaeotectonic reconstructions.

— The clayey reservoirs of the Lower Permian hold prospects in formations where the underlying deposits are oil-bearing.

— Besides the weakened zones at the contacts of textures of different types, the tectonic fractures appearing during the active disjunctive tectonic restructuring of the territory of their distribution also participated in the creation of an ideal reservoir capacity for the clayey reservoirs of the Lower Permian.

The characteristics of the composition and texture of the rocks of the Lower Permian, with their absence of silica and low amounts of OM, compel us to suggest that they could be reservoirs with indications of medium reservoir prospects.

*

4
Characteristics of Development of the Reservoir Potential of Clayey Rocks

4.1 Development of the Reservoir Potential of Clayey Rocks

The reservoir and filtration properties of clayey reservoirs have been studied in detail by various methods up to the present time [13-16, 21, 40]. However, the results obtained are so highly disparate as to preclude a correct view of the reservoir potential of these rocks. Tests carried out at the wells reveal a wide scatter of data. Not only is there lack of accord in the results of studies on reservoir parameters and the relation between yields and the determined parameters, but also conspicuous disagreement among the various investigators regarding the nature of the reservoir capacity of these rocks. Preference is given either to the porosity of the matrix, or to the fracturing, or to the parallel bedding.

To estimate the role of pores in the development of the reservoir volume of clayey reservoirs and to obtain the quantitative characteristics of microporosity, SEM photographs were analysed over the Quantimet-720 according to the method of R.A. Konysheva and A.P. Poznikova described in Chapter 1 and were found to have been miscalculated. The distribution of the chords of pores according to their size (see Figs. 4, 5) and a comparison of pore sizes in zones with different yields (see Table 1) for the reservoirs of the Bazhenovian formation indicate that neither the quantity nor the size of the pores determines the total volume of the clayey reservoirs as reflected in the yields obtained from these rocks.

Given the situation that the properties of clayey reservoirs form as a result of the interaction of the major geochemical components of the rocks [19] but that the scale of this interaction is determined by the textural element of the rocks at the meso- and microlevels, we posed the problem of assessing the influence of the textures of different rocks on the reservoir properties of the clayey reservoirs.

Clayey reservoirs exhibit polymineral and multigenetic formations in which, alongside clayey minerals, authigenic and allothogenic minerals as well as OM are constantly present. Silica and OM often occur in rock-forming proportions. Still another characteristic of clayey reservoirs is that their component parts do not appear all at once in the sediments, which accounts not only for their compositional, but also their textural diversity.

The spatial distribution of the deposits in the basins of sedimentation is determined by a series of factors, of which the most important are the character of the tectonic dissection of the source area, composition and conditions of weathering of the rocks in the catchment area, the type of sources of the allothogenic material and relief of the basin floor, the presence of currents, waves etc. A decisive role is played by the deep-seated faults and also changes induced by the thickness of the different members of the deposits and the nature of local structures favouring the en echelon disposition of the local uplifts, their asymmetry and other characteristics. The tectonic regime is also evidenced in the structures and rates of sedimentation.

The accumulating sediments present their own mineralogical and textural imbalance in the system. The redistribution of particles of sediments and the chemical reactions accommodating the mineral components of various kinds amongst one another and also in the surrounding environment, accompanied by the solution of unstable compounds in the given physico-chemical situation and the formation of stable minerals from the products of their dissociation, are the processes that took place during diagenesis. The diagenetic stages of transformation of clayey sediments constitute an extended process that commenced soon after the formation of the first portions of matter and concluded with the lithification of the sediments, i.e., the formation of sedimentary rocks.

The initial stage of diagenetic transformation of clayey sediments was characterised by considerable flooding of the latter because differences in the composition and concentrations of salts in the pore waters and the surface waters was practically absent. Precisely at this time, silica, actively consumed by siliceous planktonic organisms, started reaching the basin water. Later, processes leading to the loss of water and enrichment of silica gave rise to a situation wherein the saturation limit was exceeded and the waters became supersaturated, resulting in the deposition of silica or silicic acid along with silicification of the clayey and organic portions of the rocks.

One of the important characteristics of the clay minerals is their capacity for ion-exchange, expressed as the rate of exchange capacity. In the self-same clay minerals, this capacity depends upon the completion of their crystalline structure. Intensive accumulation of OM in the sediments is facilitated by the widely known characteristic of clayey and carbonate minerals (in fine fraction) to adsorb organic cations from solutions, and the exchange of cations in them situated in their exchangeable positions. Exchange reactions of the inorganic complex in the organic took place very quickly because the cations situated on the outer surface of the clay minerals underwent exchange in the organic distinctly faster than the interlayered cations. Because of the rapidity of the ion-exchange reactions of the clay minerals reaching the sediments from the source area, equilibrium was readily obtained with the cations of the marine waters; hence the clay minerals of the sediments accommodated no additional OM other than that already present.

Silica precipitating from the solution was adsorbed on the surface of the microblocks of clay minerals both with the adsorbed OM and without it. It was also adsorbed by the fine granular carbonate minerals being formed in the basin of sedimentation and was incorporated in the sediments in a weakly crystalline state. In the sediments from which clayey reservoirs of the Bazhenovian type later formed, silica developing as coatings on the microblocks of clay minerals and cementing them to each other, provided the skeletal framework within which the clay minerals retained significant moisture and thereby plasticity. The uncemented parts of silica assisted in the adsorption of additional portions of OM, which hydrophobised not only the free surface of the clayey microblocks, but also the surface of the silica coatings [envelopes].

The adsorption of silica by the surfaces of the minerals composing the rocks of the Domanikian type led to the formation of a skeletal framework that is more or less uniform only sporadically in the rocks and primarily in those parts where the clay minerals and fine-grained carbonate materials are situated. Hence, the weakened zones appearing at the contacts of the silicified parts of rocks of such a type might extend for large distances but overall would be found to be less than in the rocks of the Bazhenovian formation and Lower Menilitovian member.

The deposition of the main mass of silica in the rocks of the Khadumian and Batalpashinian formations took place at the late stage of diagenetic transformation of the sediments, when they had practically lost all their interstitial waters, which remained only in places where the clay minerals were deprived of the adsorbed OM. It is precisely these parts, which possessed a highly bizarre form, that proved unique for silica precipitation. The deposition of silica at these places rendered them more brittle and it is here that the dislocations which form the major part of the general volume of these rocks appear. The separation of the rocks along the zone of dislocation was accompanied by the opening up of pores situated along the path of dislocation. The dimensions of the widening are not as significant as in the Domanikian horizon although the increase in volume amounts to 12 to 15% of the total capacity, creating the weakened zones in the dislocated parts. The microtextures played a distinctly subordinate role and formed at the contacts of minerals with silica envelopes and with the hydrophobic shell formed by the OM, and also at the contacts of the clay minerals with pyroclastic or terrigenous particles constituting the centres of stresses.

The primary textures appearing in sediments as a result of distribution in space of the minerals and organic components underwent considerable modification during diagenesis under the influence of specific conditions related to the arrival of silica in the sediments at different times. The textural heterogeneity increased, which at the first instance influenced the changes of microtextures and the formation at their contacts of additional volumes of weakened zones. Changes occurred in the affected part of the mesotextures, as a result of which specific weakened zone dislocations appeared in the rocks of the Khadumian and Batalpashinian formation (see earlier discussion).

It is necessary to note one very important detail of the post-depositional modification of the textural aspect of the sediments to which the clayey reservoirs were later subjected. This is the fact that the silica that migrated at the end of diagenesis was not yet related to the organisms but incorporated in the sediments directly, thereby increasing the temperature of transformation of the sediments into rock. The second stage of silicification preceded the transformation of the OM adsorbed by the clay and carbonate minerals at various stages of the sedimentary process, and silica replacement took place until the end of diagenesis. The sediments were converted into rocks. The increased temperature favouring the second stage of silicification promoted the catalytic transformation of the adsorbed OM due to which a peculiar redistribution of OM and hydrophobised surface of the clayey and other mineral components resulted. The mobile components formed during the transformation of OM hydrophobised the surface of clayey and fine-grained carbonate minerals and also the silica adsorbed partly by the minerals and organic components.

Repetition of the process of hydrophobisation of the constituent parts of the deposits is characteristic of clayey reservoirs. The primary hydrophobisation of the clayey and parts of the carbonate minerals took place during sedimentation. During early diagenetic silicification, part of the hydrophobised minerals remained unaltered but part was covered by a film of silica. Some of the products of transformation of OM were adsorbed by the minerals, but some by the newly formed silica. Complex aggregates formed, comprising not only clayey or carbonate matter but also adsorbed organic products, silica and OM. In certain parts of the rocks, some members comprising either adsorbed silica only or a single layer of OM were formed.

The contact of all these disconnected components created a textural heterogeneity, along whose boundaries the weakened zones formed. The reservoir capacity of clayey reservoirs is contributed by pores of different forms and zones of contacts of meso- and microtextures, amongst which the mentioned complex mineral-siliceous-organic aggregates played a significant role.

The most complex multilayered aggregates, composed of alternating clay minerals, OM and silica, display distinct development in the rocks of the Bazhenovian formation because there is a considerable quantity of clayey matter (from 50 to 65%) in these rocks. Hence their reservoir capacity was formed under the influence of the weakened zones at the microlevel. Thus the length of the weakened zones at the meso- and microlevel falling within unity of a miscalculated field and measured for samples from well 118 of the Salym oil-field is 5.25 and 70.75 mm/mm^2. The coefficient of orientation is 1.446 and 1.623. Coefficients of orientation measured over samples from well 49 correspond to 1.966 and 1.693, which supports the primary development of the weakened zones perpendicular to the bedding. This is related to the condition that the textural heterogeneity obtained its final shape as a result of silicification at the late stage of diagenesis, the material for which reached the sediments from the bottom. A similar mechanism of formation of the reservoir capacity is invoked for the rocks of the lower Menilitov member, a potential reservoir of oil of the type found in the Bazhenovian formation.

In the rocks of the Domanikian horizon, whose textural inhomogeneity was produced as a result of the interaction of four rock-forming components (the most important amongst them are the chemogenic and shell carbonates), the mesotextures and their contacts were vital to the formation of the reservoir capacity, followed by the microtextures, which played an additional role. Hydrophobisation of the clayey and carbonate minerals in these rocks, as shown in Chapter 1.3, likewise did not take place at one time. According to this mechanism the weakened zones were formed in the rocks of the Kuonamian and Inikanian formations. The mean size of the chords of pores in the rocks of the Kuonamian formation varies from 3.7 to 3.8 μm parallel to the bedding and from 2.7 to 2.9 μm perpendicular to the bedding. The absolute length of the weakened zones at the mesolevel changes from 250 to 348 μm parallel to the bedding and from 368 to 426 μm perpendicular to the bedding. The coefficient of orientation varies from 1.244 to 1.467.

The mechanism of formation of the reservoir space in the rocks of the Khadumian formation has already been examined. Some data are given here. The average size of the chords of the pores is 0.80–1.5 μm parallel to the bedding and 0.63–1.02 μm perpendicular to it. The coefficient of orientation varies from 1.543 to 1.605.

The zones of contacts of texturally different parts constitute the major areas of reservoir space in the clayey reservoirs and the chief paths along which oil migrates through these rocks and disengages these zones so as to form oil-pools in the clayey reservoirs, and is also diverted to the bottom of oil-wells during the tapping of oil in such a reservoir. Having been formed during sedimentation and assuming final spatial configuration in diagenesis, the zones of contacts require very large forces to keep them wide open. This is possible only during the vertical migration of oil, which is carried out as a double action–opening up the weakened zones and forming commercial pools of oil in them. After the withdrawal of oil the fractures close again and their opening now becomes impossible. This takes place during the lowering of formational pressure below the hydrostatic at the time of the hydraulic operation, after which the weakened zones close and the impulses simply fade away.

4.2 Conditions of Migration of Oil and Gas in Clayey Reservoirs

In most models of formation of oil and gas reserves, the migration of hydrocarbons (HC) from their places of formation in the rocks to the reservoirs is suggested. Considerable significance is attached to the characteristics of migration through different types of rocks situated in the paths of migration of HC. The filtration characteristics of liquids and gases through sandy silt and carbonate rocks have been studied over many real and experimental materials. The processes of migration of HC through clayey rocks have been investigated through a few experiments. It was established that these processes are difficult but not impossible. As shown by V.P. Savchenko in 1958, in order for oil or gas to pass through water-saturated rock with capillary or subcapillary structure of pore space, they must be subjected to a certain excess pressure, termed the injection pressure. The injection pressure under which gas or oil penetrates through such a rock is called the 'break-through pressure'.

The break-through pressure, as shown by our experiments, is variable and depends on the structural-textural characteristics of the clayey rock series through which migration takes place. The textural heterogeneity characteristic of clayey rocks favours the presence in them of parts with sharply differing permeability. The migration of HC takes place along these parts, i.e., the weakened zones. The residual traces of HC left behind during their movement are easily determined during mineralogical-petrographic investigations of clayey rocks with different textures.

Clayey reservoirs possess the characteristic of not only transmitting the HC through them, but also accumulating them in channels which form during the widening of the weakened zones at the time of migration. This is possible because of the textural inhomogeneity of the clayey rocks, which are also characterised by the hydrophobisation of OM and silicification of the zones of contacts of different textural parts. The greater the hydrophobisation of the contacts in the clayey reservoirs, the larger the width of the possible widening of the weakened zones and the greater the content of oil or gas emplaced.

Our studies have shown beyond doubt that all the clayey reservoirs were formed at the moment of migration of oil in them in the tectonically active rock parts. The presence of such tectonically active rock parts in the Salym oil-field is demonstrated by a series of parameters. For example, an analysis of the Bazhenovian rocks (Fig. 42) helps to divide the oil-bearing formations into at least two separate tectonically set blocks–the northern and the southern. The structural trend within the limits of these blocks, obtained from an analysis of the zones of lower thicknesses of the formations, is, interestingly, reversed. In the northern block the layer Yu_0 possesses a north-west trend while in the southern block the trend is north-east [30]. Such a sharp change in trend ought to indicate a fault zone boundary, which, according to the data of MOGT, is distinctly expressed on the surface of the pre-Jurassic deposits and is clearly

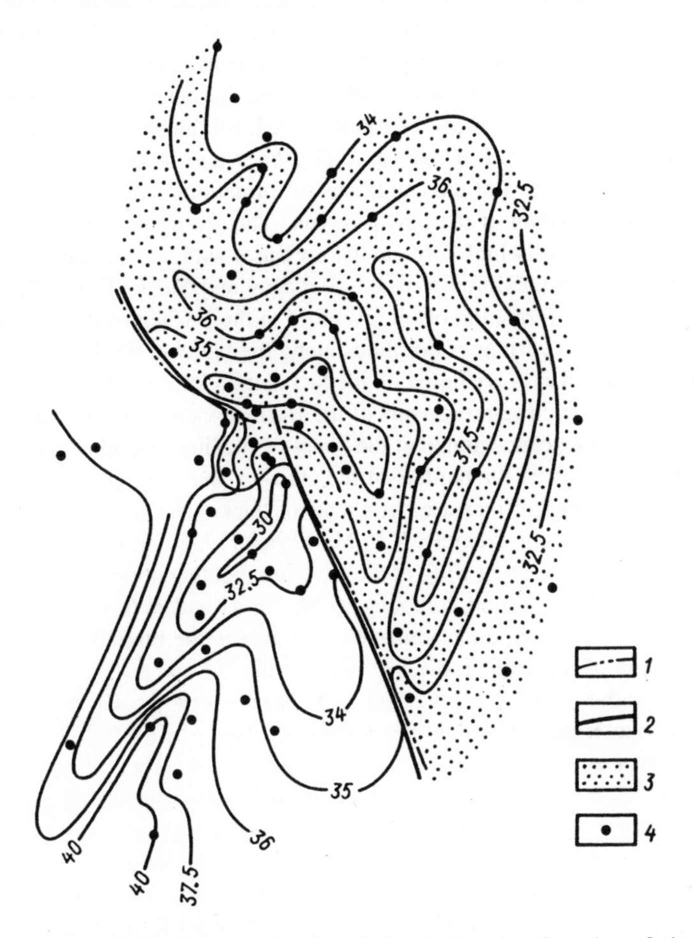

Fig. 42. Scheme of isopachs (m) of rocks of the Bazhenovian formation of the Salym oil-field, western Siberia (after L.P. Klimushina, 1980).

seen intersecting the entire sequence of the Neocomian deposits. This belt precisely serves as the eastern boundary of the distribution of the members of the Achimovian sandstones (tilted horizon d_2, according to MOGT) and the western boundary of the sandstones of the layer BS_6 becoming clayey southwards.

The exceedingly narrow belt of decreased thickness at the southern block with a north-east trend, belonging to the arched part of the recent uplift, was also caused by the sheared and faulted nature of the rocks. The materials of the MOGT, the sharp temperature anomaly and attenuation of the rocks exemplify the presence in this zone of faulting in the pre-Jurassic and Jurassic deposits.

Wells with increased oil yields from the layer Yu_0 are confined to the boundary of the active zone of migration (fault zones). The zones of increased pressure and temperature are also related to such a belt. The presence of faults is an undoubted condition of oil formation in the Bazhenovian formation. This point of view is held by many researchers (K.I. Mikulenko, A.I. Stepanov and Yu. A. Tereshchenko, K.F. Udalova, U.G. Ishaev and others), who have analysed the various aspects of the conditions of formation of oil in the rocks of the Bazhenovian formation of the Salym region. The spatial blending of the zones of AHFP (anomalously high formational pressure) with linear anomalies of temperature and oil outputs prompted A.I. Stepanov and Yu. A. Tereshchenko [39] to suggest that the formation of oil resources in the Bazhenovian formation was associated with the current of hot fluids along the faults in the Jurassic deposits. All the foregoing views are based on the fact that all known oil-pools and occurrences of oil in the clayey reservoirs of the Bazhenovian formation are restricted to the local uplifts, concentrated at the zones of maximum gradients of the thick Neocomian deposits at the borders of the Ob'-Pur and Koltogor troughs. At the Salym uplift the belt of increased gradients of thicknesses is extended, as established by scientists, for example, V.P. Markevich (1966), almost in a meridional direction along the western wing of the structure, holding onto its crestal part and exposing beyond its closed contours. G.M. Taruts and E.A. Gaideburova [41] have drawn attention to the leading role of faults in the migration of oil and in the formation of oil-pools in the Salym oil-field.

B.V. Kornev, M.I. Kozlova and L.A. Bedenko [30] used the method of bringing out tensional faults developed in 1967 by V.P. Markevich and M.I. Kozlova and constructed a structural map of the roof of the formation Yu_0 of the Salym oil-field in which the tensional faults intersecting the structure in a submeridional direction were located. Three blocks were distinguished as divided by tensional faults: the western, the central and the eastern. The central block is the most uplifted, wherein wells with maximum yields are situated. The authors observed that the structures of the central Ob' River region, in which the oil-bearing rocks of the layer Yu_0 are situated, are immediately contiguous to the regional faults of the Ob'-Pur trough.

E.G. Kovalenko [30] utilised the structural geomorphological method of exposing the zones of increased fracturing of the earth's crust, which appear in an orderly fashion in the relief and the landscape in the form of lineaments, and established for the territory of Great Salym transregional, submeridional and sublatitudinal systems as well as fracture systems trending north-west and north-east extending for 50–100 km and more, related to deep-seated faults. The lineament systems of much smaller fractures are associated with the mobility of the individual blocks of the basement at the later and recent stages. An analysis of lineaments and the occurrence of oil in the Bazhenovian formation, in the opinion of E.G. Kovalenko, indicate the presence of wells of increased output in the zones which belong to the extended system of lineaments and particularly the intersections of the 'tectonic network,' which are the most weakened zones, with increased reservoir volume properties of the rocks of the Bazhenovian formation and the development of AHFP.

An analysis of the magnetic field carried out by V.S. Surkov and his co-workers in 1974 prompted them to conclude that those parts of the layer Yu_0 with increased productivity are related to the occurrence of faults in the basement. A comparison of the oil occurrence in the Bazhenovian formation with the character of magnetic anomaly similarly led them to conclude that a considerable part of the productive wells is situated in the gradient zones of magnetic anomaly or is confined to the chain of local positive anomalies lying sandwiched in the field with other trends of the elements. However, in the northern cupola of the Lempin and in the Northern Salym uplifts, the productive wells are situated almost at the centre of the minima of the negative magnetic field.

According to the data from a highly accurate gravimetric survey conducted by I.I. Vernik and A.I. Chashchin in 1972, two zones of local minima of gravity anomalies are distinguishable in the territory of the Salym oil-field. Based on these minima, in 1982, V.A. Boldyrev, V.I. Blyumentsvaig and N.D. Kantor presented a prognosis of the parts of increased fracturing which coincide with parts of increased porosity of the Bazhenovian rocks and increased productivity of the layer Yu_0. The high-accuracy gravimetric data support the fact that the differentiated zones of minima are connected not only with zones of increased fracturing in the rocks of the Bazhenovian formation, but also with the zones of attenuation in the lower formations, i.e., indicate a connection between the oil occurrence in the Bazhenovian formation and the structure of the rock complexes of the lower formations.

A comparison of the fields of high-fidelity gravity and magnetic data contours with the results of thickness obtained, the reservoir properties and the oil occurrence in the rocks of the Bazhenovian formation, led V.A. Boldyrev and his colleagues to conclude that oil occurs in the Bazhenovian formation either close to the inferred basement faults or within the limits of the active blocks of the basement. Based on magnetometric data, sublatitudinal or submeridional faults were located in the Salym oil-field. The existence of some of these faults has been confirmed by geological and seismic data.

The following characterisation of the productivity of the Bazhenovian formation was given by N.A. Trapeznikova, A.A. Kharlanova, S.V. Zui and T.A. Bugrimova in 1986, based on their analysis of the data of MOGT:

— Those parts where high-yield wells are situated are confined to the structural discontinuity of the seismic horizon B.

— High productive zones are normally accompanied by a conspicuous deterioration of the reflection characteristics of the rocks and a specific wave pattern (change of seismofacies) of the horizons of the Tyumenian formation and the base of the sedimentary cover; in particular, a distinct peculiarity of such parts is noticed for the time gaps of short amplitude.

— Those parts in which dry wells are located normally correspond to a good reflection characteristic of the deeper lying deposits of the Tyumenian formation and better quality of reflection of the base of the sedimentary cover; to these parts in the superimposed Achimovian deposits belong the thin layers of sandstones with high acoustical impedance, which increases the intensity of reflection of the B horizon and lessens its apparent frequency.

The above-mentioned and many other investigators have employed various methods for mapping the faults in the rocks of the Bazhenovian formation and shown that the presence of zones of stress was favourable for the migration of fluids. Based on our data from analytical studies of pore space in the rocks of the Bazhenovian formation (see Table 1 and Figs. 4, 5), the presence of a stressed zone is fixed; Pore space is thrice larger in the zone of tension than in the limb region.

Studies on the complexes of microfossils from the oils of the Bazhenovian formation of the Salym and Malobalyk oil-fields were undertaken by A.M. Medvedeva in order to determine the role and character of the migration processes in the formation of the oil reserves in the rocks of the Bazhenovian formation [30]. It was established that each complex contains a complex collection of microfossils comprising *'in situ'* spores and dusts of Jurassic age and 'migrated' Palaeozoic spores. Acritarchs are present in considerable quantity in all the oil samples of the Salym and Malobalyk oil-fields. The largest quantity of microfossils was encountered in oils from wells 42 (2845–2885 m) and 32 (2745–2785 m) of the Salym field. But the proportions of *in situ* and migrated parts of the complexes differ in them. In the oil samples from well 32 (at the crest), the main places contained Jurassic spores and pollen dust (34%); the migratory parts were composed of Carboniferous spores (8%) and acritarchs of the Upper Palaeozoic (16%). In the oil samples from the well at the steep limb (well 42), Jurassic spores and pollen (40%) and Upper Palaeozoic (23%) and Lower Palaeozoic (2%) acritarchs were obtained. The complexes of microfossils from the oils of the Bazhenovian formation of the Malobalyk oil reserve (well 5, depth 2879–2895 m) showed Jurassic spores and pollen (42%), Carboniferous spores (2%) and Lower Palaeozoic acritarchs (28%).

The samples of oil from the wells on the steep limb (well 42) and the crest (well 32) of the Salym oil-field were characterised by good saturation of the compounds and fairly high content of migrated spores and acritarchs. The minimal quantity of migratory microfossils of Palaeozoic age was noted in the oil samples from the wells on the limbs (wells 38, 56). It was further observed that in the oils of the Salym oil formation the Upper Palaeozoic forms were predominant, whereas in those of the Malobalyk the Lower Palaeozoic forms were predominant.

The composition of the microfossils of the oils from the Bazhenovian formation supports a wide vertical migration of the fields from the Jurassic deposits and does not justify the conclusion of the syngenetic origin of the oil present in the Bazhenovian formation. The conclusion relating to the formation of oil reserves in the Salym oil-fields due to vertical migration had already been reached in 1967. I.A. Yurkevich studied the composition of the oils from six productive horizons intersected at well 1 of the Salym field: the crust of weathering-Palaeozoic, Tyumenian formation and four productive layers belonging to the Neocomian. From the bottom upwards to the layer BC_6 the oils were characterised by the same composition, which supports the presence of vertical migration. Much later and at a different instrumental level, the presence of a wide vertical migration was likewise concluded by L.P. Klimushin and A.N. Gusev based on the results of analysis of geological material and the distribution of oils of varied compositions in the oil-pools of the Salym oil-field.

All known reserves in the Domanikian horizon of the Volga- Ural'sk region are also confined to the zones of faults and flexures, accompanied by linear relations of the fractures. Local uplifts are complicated by thrust dislocations. Two systems of tectonic shears find expression in the longitudinal and transverse faults, which are traced along the trends of the faults of the basement and terrigenous Devonian [33].

Based on the results of studies on the complexes of vegetal microremains isolated from the rocks, the bitumens and oils of the Domanikian horizon, A.M. Medvedeva established that the Domanikian rocks, irrespective of their lithological facies characteristics, contain spores of Domanikian age. A part of the samples contained a considerable quantity of inclusions in the form of bituminous scum. Palaeophytological analysis conducted separately for the Domanikian rocks and for the bitumen dispersed in them, showed the presence of one and the same Domanikian spores in both. This supports the syngenetic character of the bitumen in country rocks. At the same time, the complexes of vegetal microremains isolated from oils in the reservoirs of the Domanikian horizon distinctly differ from the complexes of the vegetal remains separated from the Domanikian rocks. In all the samples of oils, in addition to the typical Domanikian forms, spores and acritarchs from the lower formations are present. This very much supports the vertical migration of hydrocarbons (HC) during the formation of the oil reserves in Domanik.

The Eastern Stavropol' basin is oriented north-west–south-east. In the same direction the basement exhibits a plunge at a depth of 3.5 to 5 km. The boundaries of the basin are complicated by faults which, in the sedimentary complex, appear in the form of diverse gravity-thrust faults and flexures.

The zone of commercial oil reserves in the clayey reservoirs of the Khadumian and Batalpashinian formations in the Zhurav structure is characterised by a distinct connection with the basement faults. As shown by P.S. Naryzhnyi (1986), all highly productive wells form narrow, linearly extended, sublatitudinal and submeridional zones.

Seismic prospecting studies (MOGT) over a large part of the Zhurav oil-field revealed the block structure of the basement. In the structural scheme along the reflected surface of the Palaeozoic, faults have been distinguished which intersect the surface in sublatitudinal and submeridional directions (see Fig. 30). It is important that the seismic prospecting works of earlier years were undertaken solely for the purpose of determining the local structures and were not directed towards tracing the faults.

During the exploratory prospecting/drilling geologists of the Stavropol' Oil and Gas Commission (P.S. Naryzhnyi and others) utilised the critical sections of seismic prospecting of MOGT, in which they differentiated the zones of loss of correlation in the Maikopian and more ancient deposits, which, in our opinion, are related to the basement faults. To verify this method, it was recommended that five exploratory wells be drilled with an average depth of 2400 m. Drilling of just one well yielded copious amounts of oil and it served as an experimental well in the exploitation operations.

Analysis of geological-geophysical material over the Zhurav oil-field led P.S. Naryzhnyi to conclude that the oil reserve in the lower Maikopian reservoirs is associated with the zones of AHFP, beyond the limits of which oil does not occur. Oil-bearing zones of AHFP are confined to tectonic faults. The width of the oil-bearing zone of AHFP is the same in different parts of the oil-field and varies from 300 m to 3 km. Oil output depends upon the location of the wells in relation to the faults, i.e., to the attenuated zones: the highest oil yield (up to 114 m^3/day) was obtained from well 62 situated immediately near the fault.

The zone of development of rocks with high productivity of oil in the Zhurav oil-field is characterised by increased tectonic activity and a change in direction of tectonic movements. The fields of development of industrial reservoirs of oil in the clayey rocks are confined to the flexural folds and over thrust zones, because of which a sharp change in the hypsometric position of the formations and the development of horst graben or gravity thrust fault structures take place. The confinement of productive zones of oil to the zones of fault tectonics supports the formation of these oil reserves in the clayey reservoirs of the Khadumian and Batalpashinian formations due to vertical migration.

The Cis-Carpathian foredeep, where the Oligocene deposits (Menilitovian formation) form potential clayey reservoirs, was formed as a result of some phases of tectonic movements. The rocks of the Menilitovian formation belong to the Cretaceous-Palaeogene structural tectonic stage, which reflects an exceptional stage of development of the Carpathian flysch geosyncline. According to the results of geophysical and deep borehole studies, transverse basins and depressions are developed in the inner and outer zones of the basement of the Cis-Carpathian foredeep which are bound by deep-seated faults. All the faults are developed in the preflysch foundation and attenuate towards the surface [29].

In the flysch complex of rocks the faults appear in the form of a closely spaced network of gravity faults, gravity overthrusts and rarely thrusts. This is exactly the character of disjunctive faults which warrant vertical migration. The characteristics of chemical composition of the oils of the Oligocene, Eocene and Upper Cretaceous deposits of the Cis-Carpathian foredeep from depths of 200 m to 5 km were studied by A.A. Orlov, V.L. Pluzhnikova, S.L. Balmasova and I.P. Ninevskii and revealed the presence of vertical movements. The ratio of pristane to phytane lies within the limits of 2–3, while the average value is 2.64, and is not dependent upon the age of the enclosing host deposits or their depth of occurrence. This supports the formation of oil in a single sedimentary basin from a single type of sapropelic-humic OM formed in weak reducing conditions of diagenesis. A majority of the oil reserves bear a resemblance in nature of distribution of H-alkanes and isoprenoids C_{10}-C_{20}. The characteristic of the quantitative distribution of H-alkanes is their maximum concentration in the field of elutriation of hydrocarbons C_{17}-C_{18}. These indications support the view that in all the reservoirs of the Inner zone of the foredeep a single genetic type of oil only is encountered.

A comparison of the oils was carried out based on the ratio of the major classes of hydrocarbons–alkanes–cyclanes–arenes according to their individual bonds and physico-chemical properties. In the benzene fraction of oil of the study region, alkanes predominate and among them is hydrocarbon of normal structure. The ratio of alkanes/cyclanes lies within the limits 1.12–2.43, H-alkanes/isoalkanes 1.23–2.25 and arenes/alkanes 0.08–0.36. Among the cyclanes, the cyclohexanes are predominant over cyclopentanes, the ratio between them being 1.2–2.36. The data obtained exemplify the presence of wide vertical migration of HC within the limits of the Inner zone of the Cis-Carpathian foredeep.

As shown by the data obtained on the features of tectonic structure of the territory, migration has been realised in a fairly narrow zone, appearing as a result of the collapse of the folds of the Palaeogene and Cretaceous flysch deposits during their overthrust on the platform and the formation of multistage recumbent anticlinal folds broken by faults at the places of formational bends. An indirect indication of the formation of oil reserves as a result of vertical migration has been established by M.P. Gabinet and L.M. Gabinet. They have pointed out that during the concentration of OM higher than 3% in the Oligocene deposits up to a depth of 4 km the impulse of active generation of HC from OM is not recognised. For rocks of these depths in which normally the deposits of the Lower Menilitovian member (potential clayey reservoirs) are situated, the stage of katagenesis of OM has been determined by $MK_1 - MK_3$. This corresponds to the stage of katagenesis of OM established from data of petrographic studies by G.M. Parparova and others [17].

In the parts bound by faults, rocks with specific mineralogical and textural characteristics stand out as clayey reservoirs of oil.

The Cambrian epicontinental basin at the beginning of the Lenian period was divided into two parts of a narrow belt of shallow-water barrier reefs, which formed and survived for a long time because of the presence along their trend of a system of deep-seated faults. Eastwards from this zone, the basin very very slowly subsided and filled with eroded material from the dry land mass consisting of clayey material and chemical precipitates of carbonates. At the same time, along the system of faults silica and other chemical elements conducive to the bloom of organic life reached the sedimentary basin. Deposits of the Kuonamian and Inikanian formations accumulated in this western arm of the epicontinental basin [15].

In such regions where the deposits of the Kuonamian and Inikanian formations have undergone subsidence to a considerable depth (in the central part of the Vilyui syneclise, according to the data of A.E. Kontorovich and others, these formations are situated at depths of 5–6 km), it should be possible to discover an accumulation of HC in them. It is essential therefore not to lose them during drilling and hence it is desirable to undertake drilling solely for the purpose of striking them and carrying out sampling.

In conclusion, while examining materials to study the migration of HC in clayey reservoirs, it is necessary to emphasise that the characteristics of clayey rocks as commercial and potential oil reservoirs were developed during vertical migration. The latter process is set in motion by disjunctive faults of the rocks of the basement extending into the overlying cover deposits, and also flexural, overthrust and other conducive deformations of the rocks and related structures in which the clayey rocks participate as potential reservoirs of oil and gas.

It has already been pointed out that the formation of clayey reservoirs and the migration of oil and gas in them constitute a unique process. The migration of oil taking place under the influence of tectonic stresses leads to the accumulation of dispersed oil in the clayey rocks along the weakened zones, and the formation of oil-pools in them. The weakened zones are separated out under the influence of tectonic stresses, which strengthen the high temperature reached through the process of injection of fluids in the rocks during tectonic restructuring of the territory.

5

Formation of Oil and Gas Reserves in Clayey Reservoirs

For the formation of oil and gas reserves in clayey reservoirs, it is imperative that the following favourable structural, tectonic and lithological-palaeogeographical conditions be present: suitable geological structure of part of the earth's crust, favourable conditions for the formation of clayey reservoirs, accumulation of oil and gas in the reservoirs and their isolation from the overlying deposits, i.e., the presence of a cover of high quality.

The preceding analyses of the geological situation in the oil and gas formations in clayey reservoirs in the USSR and other countries enable a presentation of the history of the tectonic development of these parts of the earth's crust and the formation and entrapment of the reserves in clayey reservoirs.

As is well known, the oil-bearing character of the geosynclinal fields is related to intermontane basins and piedmont foredeeps and also to the border and inner basins of platform areas divided by crestal uplifts if the thickness of the sedimentary deposits in them is high. These structures of the first order are formed as a result of the block-faulting of the basement along deep-seated faults. The chunks and blocks of varying dimensions and forms are divided by the scarps of uplift or subsidence at different levels. The transitional geostructural similarity between the geosynclinal and platformal areas is seen in the piedmont foredeeps which, in the opinion of N.A. Eremenko, nevertheless represent transitional elements between the platform and the geosyncline, albeit they are developed over a considerable part of the platformal body.

In the classification of oil- and gas-pools proposed by V.B. Olenin in 1974, based on the genesis of the structural elements and their organisation, five types are recognised. One of these types, the fourth (structures formed by tensile faults), the naming of which has not been popular, is characterised by structures of tensile origin, which are of particular interest to us, especially in solving problems relating to the formation of pools in clayey reservoirs. According to the build-up of the structural elements in this type, the oil and gas formations are divided into the following classes: (1) pre-rupture mesoclinal parts and folds, (2) pre-rupture fractured parts and (3) horsts. These classes are characterised by a typical collection of reservoirs. In the first class the reservoirs are screened along the fissures and folds, in the second along the lenses of tectonic fractures and in the third by the reservoirs represented by prominences and uplifts along the boundary fissures and screened with respect to the fissure. Among the formations studied which belong to the first class are the oil- and gas-pools of the Santa Maria basin, the Fore-Appalachian, Denver and other basins in the USA, the oil-pools of the Jela Sicilian basin, the Domanikian horizon, Volga-Ural'sk region and also potential reservoirs in the deposits of the Kuonamian and Inikanian formations. The oil-pools of the Gabon subbasin in Africa and the Zhurav reserves of Eastern Cis-Caucasus belong to the second class. The oil accumulations of the San Juan basin in the USA and the Bazhenovian formation of western Siberia likewise belong to the second class. To the third type of reservoirs may be relegated three formations, which are complicated by the potential clayey reservoirs of the Lower Menilitovian member of the Cis-Carpathian foredeep.

It is, however, necessary to emphasise that the suggested classification provides for a division of the formations according to the types of reservoirs only in a general plan because very many pools in the clayey reservoirs contain oil of a combined type. The principal characteristic of reservoirs in which the multicomponent rocks constitute the reservoirs has to be duly emphasised because all these are formed as a result of disjunctive faults. The evolution of a detailed tectonic classification of oil- and gas-pools, the reservoirs of which are formed in clayey rocks etc., is a work for the future when we should be able to safely forecast the clayey reservoirs in regions of different tectonic settings.

Deep-seated faults create the formation of reservoirs in the sedimentary rock series and among them, in the clayey rock reservoirs, oil and gas might accumulate. In the platformal areas the deep-seated faults are arranged en echelon because the independent steps and blocks were pushed unevenly and partly changed their direction of movement. In the uplifted blocks, asymmetric arches formed giving rise to the formation of flexures. Depending on the nature of block differentiation of structures of the first order, structures of far lower ranks of highly different types appeared in the sedimentary cover, including anticlines, horsts and the like. Among the anticlines, parts of the folds were extended horizontally under the action of deep-seated strike-slip faults and normal strike-slip faults. These extended anticlines are situated at an angle with reference to the fault formed, as observed in the Inglewood oil-field of the Los Angeles basin, where the feather joints intersect the main fault at an acute angle. The main fault cuts across the entire western part of this basin and runs north-west–south-east. A majority of the anticlines and brachyanticlines which follow the strike of the fault are productive.

One can understand the character of dislocations appearing during tectonic stresses based on the results of experimental models that take into account the essence of the geological processes of formation of the structures and their constituent tectonic elements. In fact, such modelling was done by A.N. Bokun in 1986. Modelling of the anticlinal folds formed during various deformations and flexural foldings imposed on the rocks of the sedimentary cover during the movements of rigid blocks of the basement along vertical and hading faults is of great significance in our appraisal of the problem. It was found that cross-folds and folds of longitudinal compression and welded types were repeated and though morphologically close differed in their inner structure due to the spatial positions of their complicated rifts. Folds with cross-folding exhibited two systems of gravity-thrust faults with dip angles 75–90°C extending from the borders of the block to the crestal part. Folds of longitudinal compression were similarly dissected by two systems of faults, thrusts and overthrusts, lying at an angle of 45°C to the line joining the foot of the limb with the centre of the fold. The welded fold was complicated by just a single system of gravity-thrust dislocations at the crestal part. Each type of fold was characterised by its typical system of fractures.

In the modelling of flexural bends two systems of shear fractures were recognised–subhorizontal and subvertical–and one system of open fractures. All the three systems of fractures formed the rift zone, which depended upon the angle of dip of the fault in the rigid basement. The zone possessed a V-shaped form in the section, started from the contact of the blocks and extended towards the top. The zones of fracturing complicating the flexural fold depended upon the angle of dip of the active fault and the rate of subsidence of the rigid basement block.

From the experimental data an important observation arose, namely, the greater the thickness of the deposits separating the potential clayey reservoir from the surface of the basement, obviously the less intense its thinning. This conclusion has been cited in a monograph on the factual data published with regards to the productivity of oil reserved in clayey reservoirs both in the USSR and in other countries.

To illustrate the above conclusion, we compared the well discharges and oil yields from the Domanikian deposits of the Volga-Ural'sk region and found that the average yield lies within the limits of 60 m^3/day, whereas the yield from the rocks of the Bazhenovian formation exceeds 300 m^3/day. In the former case the clayey reservoirs are separated from the basement by a thick rock series (average around 2,000 m), whereas the productive layers of the Bazhenovian formation are separated from the rocks of the basement through a distance of up to 500 m. In the numerous linear swells and other types of uplifts, in the building of which the Domanikian deposits participate, the faults of the basement are expressed in the form of flexural folds, for which, as A.N. Bokun demonstrated in his model, a unique V-shaped zone extending upwards is characteristic. Hence, in the Domanikian deposits only the parts adjacent to the zones tracing the basement faults would prove productive.

For the anticlinal folds of the groups of oil formations of the Greater Salym comprising the rocks of the Bazhenovian formation, the welded type is characteristic and the fold is intersected by a single system of gravity-thrust faults accompanied by the zone of thinning at the crestal part of the fold.

In the folds of the Cis-Carpathian foredeep a longitudinal type of compression predominates with the formation of gravity-overthrust faults and two zones of thinning adjacent to the faults running from the base of the limbs to the centre of the folds. Such a type of folds has been described by I.V. Vysotskii in the Borislavian zone where the linear anticlinal folds are situated in the escarpments separated by overthrusts and dissected thrusts. In his opinion, multiple inversions are characteristic of the structure of the Cis-Carpathian foredeep. The terminal formation of the structures belongs to the beginning of the Miocene. To such a type belongs the Zhurav uplift where the zone of thinning is related to the system of en echelon (termed 'cascade' by I.V. Vysotskii) or stepped zones at the border of the structures of the second order of different ages and movements.

Frequent and thick inversions of tectonic movements are helpful not only in the formation but also in the break-up of the already formed oil formation. Hence, highly practical significance is attached to studies devoted not only to the tracing of faults, but what is very important, to the determination of the time of their formation and of the rejuvenation of tectonic activity.

An analysis of the tectonic setting of the structures wherein the clayey reservoirs participate both as containers of commercial reserves of hydrocarbons and potential petroliferous formations, supports the fact that for the development of their industrial potential, tectonic activation at the stages of accumulation and diagenetic transformation of the sediments and also at the final stage of formation of the oil-pools, was essential.

The basins studied in which industrial and potential clayey reservoirs have accumulated bear in them relict traces of the tectonic framework of the preceding period. Furthermore, during the process of sedimentation, structures of regeneration formed in them which favour the rugged character of the basin floor. This feature is characteristic of all other sedimentary basins where accumulated rocks capable of containing some quantity of these or other hydrocarbons (HC) exist. However, for the studied basins a rejuvenation of tectonic activity was characteristic of the zones of fault displacements, as a result of which they remained the pathways along which the components were supplied to the waters of the basin during sedimentation and to the sediments during diagenesis. This reveals their significant influence on the operation of the entire sedimentary process and in particular on the formation of the reservoir capacity of the clayey reservoirs. Such processes include the blooming of organic life, supplying to the sediments the products of its own biological activity, silicification of the rocks (silica pockets in the mineral and organic components, siliceous framework, dislocations), change of textures with the formation of weakened zones, creation of anomalous pressure and increased temperature, exerting a profound influence on the transformation of OM of sediments and rocks, and the formation of hydrophobic fragments in the clayey and fine granular carbonate minerals and also in the organic matter (OM) of the first and second types.

The adsorption of organic matter (OM) of the first and second types and the products of transformation of the OM of the third type produced around their particles a peculiar protective film, which prevented them from being linked to some part of the same reservoir when it became saturated with commercial oil. This process embraces considerable space in such OM-enriched rocks as the Bazhenovian formation, the Domanikian horizon and to a lesser degree, the Khadumian and the Batalpashinian formations. This holds equally true for the rocks, potential reservoirs of oil, of the Menilitovian, the Kuonamian and the Inikanian formations. It is precisely the property of the OM of the rocks to absorb the hydrocarbons of oil which gives an impression of the absolute or significant similarity between the hydrocarbon part of the oil and the OM part of the clayey reservoirs.

As mentioned earlier, diagenetic silicification led to the formation of rigid skeletal remains within which the clayey matter contained considerable moisture and dispersed less vigorously so that the formation of AHFP (anomalous high formational pressure) was promoted. This appeared under the influence of the increased temperature provided by the hot fluids arriving through the faults.

In the section dealing with the clayey reservoirs of the Bazhenovian formation, one view pertaining to the formation of oil from OM in the rock formation was discussed and the impossibility of formation of oil from undoubted OM was demonstrated on the basis of an analysis of the characteristics of transformation of OM at its contact with the minerals of the sedimentary rocks. The references cited by us earlier and the data of M.F. Sokolova indicate the inhibition in the transformation of montmorillonite into hydromica both because of insufficiency of cations, especially in potassium, and also because of the difference in the lattice discharges of these minerals. A.I. Stepnanov and Yu. A. Tereshchenko [39] examined this question from one more aspect. They quite correctly pointed out that during the determination of the conditions of formation of the oil reserves in the Bazhenovian formation, characterised by anomalous high formational pressures and temperatures, it is not necessary to restrict assessment solely to the Salym petroliferous formation. Leaning heavily on the data of O.G. Zaripova, V.P. Sonicha and K.S. Yusupova [30], A.I. Stepanov and Yu. A. Tereshchenko showed that in the central Ob' oil- and gas-bearing fields, where the quantity of OM, type of rocks and their depth of occurrence correspond to those of the Salym field, a majority of the wells drilled in the deposits of the Bazhenovian formation did not strike oil in spite of commercial potential indications. This convinced the authors that the indications of oil formation in the Bazhenovian from syngenetic OM are unfounded.

After studying the influence of the rate of sedimentation on the compaction of clayey and partly carbonate and siliceous deposits, L.A. Nazarkin came to a similar conclusion. He established that stratigraphic intervals with a low rate of sedimentation of clayey, sandy-clayey and carbonate rock series are characterised by low compaction of the rocks. Low rates of sedimentation led to consolidation of the links between the different particles of the rocks and between the minerals and the organic components due to both molecular and ionic bonding. The rocks of the Bazhenovian formation, the Domanikian horizon, the Maikopian formation and others belong to this type. The presence of a rigid skeleton in these rocks (incidentally, we had already written in 1977 about the presence of a rigid skeleton) supports the fact that during subsidence and subsequent leading, they experienced a pressure that exceeded their structural strength. As a result, they underwent 'stressed' consolidation which impeded the transformation of syngenetic OM into liquid hydrocarbons.

A.K. Mal'tseva and N.A. Krylov (1986) in their work on the Jurassic formations of the Central European platform and the occurrence of oil and gas in them, analysed the influence of vertical tectonic movements on the character of cyclicity and concluded that the nature of the cyclic development of the epi-Palaeozoic plate in the Jurassic times was non-synchronous and that the individual phases of the Jurassic cycle of the second order in various plates did not conform in time. For the Western Siberian plate, an important consequence of this conclusion is the fact that the phases of differentiation and regression on this plate were reduced and hence the rock series of bituminous clays of the Bazhenovian formation appearing along the boundary of the Jurassic and Cretaceous systems, never fell within the zone of oil formation.

Still, one evidence for the unsoundness of the point of view proposed above is the character of the change in formational pressure and particularly formational temperature right at the Salym field, where the clayey reservoirs contain commercial reserves of oil. Here, the formational temperature increases from the limbs of the structure to the crestal part and the zones of increased temperature possess a linear trend close to the meridional. From this a conclusion [39] may be drawn, namely, that such a distinct and systematic change of the formational temperature cannot be related solely to the local processes of oil formation.

A similar conclusion has been presented by A.R. Kurchikova and V.P. Stavitskii [40] who hold the view that local warming up of the rocks of the Bazhenovian formation is supported by factual material in favour of the fact that for the roof rocks of this formation a large quantity of areally varying local maxima and minima of temperatures is characteristic. Their particularly high incidence in the southern and central regions of western Siberia and the limited extremes in the north could possibly be attributed to an insufficient study of the problem.

The local temperature maxima are observed in the Salym (excess over the background value reaches 40°C), Krasnolenin (up to 20°C), Vartov (up to 10°C),

Pudin (up to 20°C), Mezhov (up to 10°C), Nadym (10–15°C), Gubkin (up to 20°C) and other regions. The local temperature minima are fixed in the Surgut (up to 20°C), Vengapur (up to 10°C), Dem'yan (up to 10°C) and other regions. These anomalies are related to an increase or decrease in depth of occurrence of the Bazhenovian formation. Thus, within the limits of the Lempin and Malosalym fields in a section at a depth of 2000 m, the temperature increases generally to 90–95°C; for the Salym oil-field it has been established by and large that the higher the temperature of the rocks of the Bazhenovian formation, the greater it was at the 2000 m level. In the relatively less heated parts of the Pravdin oil formation, the temperature at a depth of 2000 m varies only between 70 and 75°C.

Hence, the anomalous high temperature in some parts of the Salym petroliferous formation is not confined to separate intervals and where it exists, is characteristic of the entire section of the deposit. The source of the heat, according to A.R. Kurchikov, is the upper part of the Jurassic basement or the sedimentary cover, that is, the rocks underlying the Bazhenovian formation. Similar data on the temperature of the Tyumenian and Bazhenovian formation have been presented in the work of A.M. Brindzinskii and others[1].

The data presented bring out the distinct connection between anomalous high temperature and the basement faults reflected on the superincumbent deposits. The commercial and potential clayey reservoirs described above are similarly characterised by increased temperature (above 100°C in the Domanikian horizon, Khadumian, Batalpashinian and Menilitovian formations), with fields of distribution occurring as narrow belts along the lines of faults, and a gradual decrease in temperature occurring proportionate to distance from the fault zone.

The absence of any link between the occurrence of commercial oil and gas reserves and the OM content in clayey reservoirs is also confirmed by the results of thermographic studies on the concentrates of dispersed OM. The combined interpretation of the curves of differential thermal analysis and thermogravimetry showed that the OM present in the composition of the rocks of the Bazhenovian formation of the Salym petroliferous belt (wells 49 and 125) belongs to the first type, is composed of fusinite and semivitrinite and contains negligible characteristics to generate liquid hydrocarbons. At the same time, the OM from the rocks of the Bazhenovian formation of the Talin oil-field does not contain oil and belongs to the first type with a small admixture of OM of the third type, and hence possesses more positive characteristics for generating liquid HC than the OM of rocks with a commercial oil-bearing character. The OM from the rocks of the Khanty-Mansii field occupies an intermediate position between the two.

The OM occurring in the deposits of the Lower Maikopian of Eastern Cis-

---

[1]A.M. Brindzinskii, I.I. Nesterov, G.R. Novikov and others. 1971. Salym oil-bearing region. *Trudy ZapSibNIGNI, Tyumen*, no. 41, p. 314.

Caucasus (Zhurav oil-field, wells 62 and 67) possesses another characteristic. It belongs to the third type, consists of plant tissues with predominant vitrinite and possesses considerable potential to generate liquid HC. This OM existed in the low and middle stage of transformation, which means that its potential to generate oil has not yet been realised. Nevertheless, the rocks in which this OM is prevalent hold commercial oil reserves.

Data from studies of scattered OM together with studies of the petrographic types of OM of clayey reservoirs and characteristics of its transformation during contact with minerals of the sedimentary rocks, not only of the Bazhenovian formation, but also of all other clayey reservoirs, support the view that the genesis of oil and gas at the sole expense of OM in the rocks ought to be rejected, based on the factual data above.

The oil and gas reserves in clayey reservoirs formed as a result of vertical migration, because invariably the clayey reservoirs with commercial reserves of HC are situated in the lower structural stages of the sedimentary cover. The migration of oil and gas along the weakened zones of rocks and the formation of reserves of HC in the clayey reservoirs were the concluding acts of tectonic activation of the territory of development of the deposits—the potential clayey reservoirs.

During studies on the deposits of the Domanikian horizon and the Bazhenovian formation the investigators noticed that in the basins where the materials of the future clayey reservoirs had accumulated, an incomplete compensation of sagging due to the sediments formed, had occurred. This characteristic was determined through an analysis of palaeotectonic structures in which the thickness of the study deposits was taken into account. The results of the study of these deposits described in the previous chapters reveal that the thickness of the sediments accumulating due to the arrival of material from the continent (as in the deposits of the Bazhenovian formation) or due to the addition of allothogenic material with the generation of a chemical deposit in the basin (as in the rocks of the Domanikian horizon), is considerably less than the recent thickness of these deposits.

The entry of endogenic components in the sedimentary basin at various stages of the sedimentary process constituted a peculiar admixture to the sediments being formed. For the studied commercial and potential clayey reservoirs this marked the end of sedimentation and the beginning and middle of diagenesis.

In all the stages of formation of clayey reservoirs, the areas of their accumulation experienced tectonic stresses as a result of the formation of faults appearing at different periods, and the entry of endogenic components into the sediments increased the thickness of the deposits under formation. Silica played a distinctive role and, depending upon the time of its contribution, sometimes increased the thickness of the sediments by one-third. That led simultaneously either to intense blooming of the planktonic organisms or the formation of a silica framework protecting the sediments from intensive compaction, or to the creation of zones of future faults, in other words, in every case to the development of the needed useful space for the clayey reservoirs.

The primary development of vertical faults is decided by the magnitude of the weakened zones developed perpendicular to the bedding, which exceed 1.5–3.5 times the number of weakened zones parallel to the bedding in spite of the distinct horizontal layering characteristic of clayey rock reservoirs.

It is interesting to analyse the reasons for the low level of entry of allothogenic material from the continent into the basin and the insignificant bioproductivity and also chemical precipitate of carbonate in the corresponding basins. The chief clay mineral of the clayey reservoirs is the mica-like mineral of the hydromica type–montmorillonite (1 M)–with a poorly developed structure and a lattice ratio of 2:1. The halo-forming character of its basal reflection (1 nm) indicates a large admixture of the swelling component. This peculiar 1 nm reflection is particularly conspicuous in the Bazhenovian formation and the Lower Menilitovian member. In the rocks of the Domanikian horizon this peak shows a somewhat subdued halo. In order to explain the conditions of accumulation in the rocks of a hydromica component with such a typical 1 nm peak, it is necessary to trace its passage from the disintegrated rocks at the source to its present position in the deposits through transport in the basin of sedimentation.

Conditions of denudation of the rocks at the source areas and the formation of suspended materials and colloids of rivers depends upon the climate and ruggedness of the relief. It is well known that the intensity of denudation in the recent plain areas is 10 times less than in the montane regions. The character of ruggedness of relief also influences the velocity of river flows, which affects the granular distribution of sediments. Slow removal of the products of weathering facilitates intensification of chemical weathering and increases the role of OM in this process. As shown by I.I. Ginzburg, all magmatic and metamorphic minerals as phased products of weathering form hydromicaceous minerals. The degree of their completion depends on the intensity of chemical weathering and quantity and type of OM, that is, the depth of transformation of the parent rocks.

Clayey reservoirs, as three- and four-component bodies, do not, however, contain an admixture of terrigenous minerals; only a small quantity of the latter is present in them in the form of fine silty or coarse pelitic material, differing very little from the main allothogenic clayey part of the rock. This reflects the high intensity of chemical weathering and the long distance of the source of the load of clastic material.

In the reservoirs studied, intensive chemical weathering led to degradation of the hydromicas, which appeared in the magmatic and metamorphic minerals as their favourable energy substitutes. With the loss of potassium in the acidic environment and stimulated by OM, part of the degraded micas was converted into the mixed-layered form of hydromica, montmorillonite, with a poorly developed structure. The degraded hydromica and the mixed-layered minerals, possessing a high sorption capacity, in part formed organic mineral complexes which together with products of transformation of diverse OM participated in the processes of chemical weathering. Then these organic mineral complexes together with the hydromica group of minerals not adsorbed by the OM and with both components of the sapropelic and humic OM, reached the river system and [from there] the basins of sedimentation. The humic part of OM reached only basins where three-component clayey reservoirs formed in which the rock-forming components were primarily clayey in composition. All the foregoing observations appertain to the rocks of the Bazhenovian formation, the Lower Menilitovian member and, to a lesser degree, the rocks of the Khadumian and Batalpashinian formations.

We fully agree with the suggestion that part of the organic-clayey components reached the basin of sedimentation as a result of transformation of the volcanogenic material forming part of the source rock region, under conditions of intensive chemical weathering aided by the solutions of organic matter.

Tectonic processes, and foremost amongst them the oscillatory vertical movements of the earth's crust, are particularly important for sedimentation and facilitate macrolayering of the rocks. These movements were responsible for the formation of thick but non-rhythmic repetition of rocks of various types during the formation of the clayey reservoirs of the Domanikian horizon and the Kuonamian and Inikanian formations. Frequent interlacing of layers at the microlevel, due to fluctuating movements of the earth's crust, is quite characteristic of clayey reservoirs. As a reflection of this process, alternation along the section of rocks of various types is observed: siliceous clays are replaced by carbonate clays, siliceous limestones, with different degrees of clayey matter, by clays with almost no admixture of siliceous or carbonate material, and clays by layers of carbonate matter with no admixture, etc. In all these rocks both the content and the type of OM vary.

Depending on the intensity of tectonic movements and the entry of endogenous components, the scale of participation of the source region supplying allothogenic clayey and organic material to the basin of sedimentation, and the new generation of minerals and the development of organic life in the same basin, changed. In all stages of the geological history of the formation of the future clayey reservoirs, the allothogenic flow of material in the zone of sedimentation was insignificant along the thickness of the sediments under formation but reduced even further, sometimes very considerably, as only tectonic activity was alive. Along the faults the arrival of these components stimulated the formation of carbonates and the development of organic life. The aforementioned processes took place at different periods but were not, however, rhythmic in their repetition. Their importance varies in different clayey reservoirs. Thus, in the clayey reservoirs of the Bazhenovian type the new generation of carbonate minerals and fauna with carbonate skeleton are insignificantly developed and the organic life is represented mainly by planktonic organisms.

In the rocks of the Domanikian type, the clayey material arriving from the source region was sometimes almost completely recrystallised so that the formation of the sediments came under considerable influence of the chemical precipitation of carbonate material which, for example, as shown by S.V. Maksimov, comprised 35% of the sediments of the Domanikian period.

Thus the mineral composition of the clayey reservoirs formed through the participation of both allothogenic and authigenic components. In the rocks of the Lower Maikopian of the Eastern Cis-Caucasus, the allothogenic source of the material reaching the site of deposition dominated during the entire period of sedimentation. As in the case of the Domanikian horizon here, too, in most cases of deposition the authigenic formation of carbonates and OM of the sapropelic type predominated and the intense supply of clayey matter led to the formation of clayey layers or clayey-carbonate rocks. Throughout the remaining period, the fraction of clayey material fell within 20 to 30% of the Domanikian rocks. Incidentally, intensive introduction of the allothogenic clayey material was accompanied by the appearance in the layers of clayey rocks of humic OM, which does not characterise the rocks of the Domanikian horizon. It appears that this OM was removed together with the clayey matter from the continent because in rocks with a predominant carbonate composition humic OM is absent.

Rocks of the Bazhenovian formation occupy an intermediate place but the allothogenic clayey minerals constitute their major component, except for the time intervals when thin layers of carbonate material accumulated. While the predominant role of sapropelic OM is due to planktonic formation, there is also a considerable admixture of humic OM in the rocks of the Bazhenovian type.

As already mentioned, OM does not occur in rock-forming proportions in the rocks of the Lower Maikopian of the Eastern Cis-Caucasus. Here we cite the fact that humic OM is practically absent in those rocks where there is a predominance of allothogenic source material. It is also necessary to emphasise the insignificant role of carbonate sedimentation in the Lower Maikopian basin. An admixture of carbonate material is noticed in the Khadumian formation along the sections of Pshekh and Morozkin streams while layers of clays with carbonate matter are of smaller thickness. From the bottom to the top the quantity of clayey rocks with a carbonate admixture diminishes. Only the middle part of the Khadumian formation–the ostracod layer–is composed of marls partly changing over to clayey limestones in which the carbonate content shows a minor dominance over the clayey. The thickness of the ostracod layer changes from one to 10 m.

Periodic rejuvenation of tectonic activity played a decisive role in the post-sedimentation history of the existing clay reservoirs. It helped their diagenetic silicification and formation of complex interrelation of silica with the allothogenic and authigenic minerals reaching the basin of sedimentation. These processes were completed by the formation of textural homogeneity and weakened zones at the meso- and microlevels.

Thus the first characteristic of the process of formation of the clayey reservoirs is the presence of the specific source area in which chemical weathering dominated under the influence of solutions of organic acids. The source rock regions were situated at considerable distances from the basins of sedimentation to which they supplied hydromica minerals with a poorly developed structure and organic mineral components consisting of degraded mixed-layered minerals of the hydromica type–montmorillonite–with a considerable quantity of micaceous pockets. The second characteristic of the formation of clayey reservoirs is the multitemporal diagenetic silicification, which led to reformation of the sedimentation textures and the appearance of weakened zones. The third characteristic is the diagenetic redistribution of OM, which favoured hydrophobisation of the contacts between the components of the rocks. These characteristics totally reflect the tectonic settings, both at the source sites and the basins of sedimentation. The tectonic nature of the filtration capacity properties of the clayey reservoirs is thus beyond dispute.

Later, as mobile chemical equilibrium was established between the sedimentary and post-sedimentary constituents of sediments, and the sediments became rocks, sufficient extension of the time module in the geological sense set in, when the rocks experienced no change. This favoured low pore permeability of these rocks and the weakened zones required considerable stresses to open or widen them. Subsequently, a new tectonic activation led to the filling of the reservoirs with oil and the formation of oil-pools. If reservoir rocks were present in the underlying deposit, they were filled with hydrocarbons. The number of reservoirs in the overlying deposits filled with oil and gas depended on the quality of cover rocks over the clayey reservoirs or over any kind of overlying reservoir. In the case of a secure cover over a clayey reservoir directly over an oil pool included in the reservoir, the anomalous high formational pressure (AHFP) was retained. It was also retained when reservoirs filled with oil were present in the overlying deposits, which fulfilled the role of a clay rock cover, absent in the case under consideration.

A comparative estimate of the prospective oil- and gas-bearing character of clayey reservoirs ought to be based on the data relating to their structure and the history of the geological development of the territories of their distribution. The data on the tectonic setting of the area of sedimentation and the characteristic of the conditions of formation of clayey reservoirs of different types require mapping of the zones of distribution of the rocks which favour accumulation of oil and gas.

The three types of clayey reservoirs differentiated below are not similar with regard to their characteristics of accumulation of hydrocarbons and the formation of an oil reserve. They consist of clay minerals (about 50%) and silica and OM (about 50%). The presence of humic components in the OM, particularly of the second type, produced favourable textures at the mesolevel and helped in the isolation of the hydrolysing components hydrophobising together with OM of the third type, the surface of the microblocks and microaggregates of the clayey and other minerals. The role of silica in the production of the effective volume capacity has already been discussed. Amongst the volcanogenic components at the micro-elemental level, uranium, which is adsorbed by humic OM, is important and its presence is reflected in the curves of natural radioactivity, the maxima of which help to differentiate the oil-saturated clayey reservoirs in a stratigraphic sequence [32, 33].

For the rocks of the Bazhenovian formation silicification alone is not that important. The time of silicification, both at the very beginning and the middle of diagenesis, is especially significant. The second important condition of oil formation in the rocks of the Bazhenovian formation is the presence of uplifts to which belong the oil reserves in the clayey rock reservoirs in conjunction with the large structure of different ages and types. Amongst the studied clayey reservoirs none absolutely analogous to the Bazhenovian formation of the Salym oil belt was struck. A subtype of the Bazhenovian type of deposits has been recognised from the potential clayey reservoirs of the Lower Menilitovian member of the Oligocene of the Cis-Carpathian foredeep. Analysis of these rocks from the wells in which oil is present in the underlying deposits provides positive results because the structural traps of the Inner zone of the Cis-Carpathian foredeep were formed by multidirectional faults and were conducive to the formation of oil in the clayey reservoirs.

The following highlight the prospects in relation to oil formation in clayey reservoirs or deposits of the Domanikian horizon type. These are four-component rocks in which silica, which played a major role in the formation of the effective space, reached the rocks at the same time as in the Bazhenovian formation. Still, its role here is not that high because clay minerals constitute not more than 30% of the rocks and hence complex organic clayey-siliceous aggregates are sparsely distributed here. Silicification of the part containing fine-grained carbonate material increased the textural inhomogeneity but not to the extent to which it is observed in the reservoirs of the Bazhenovian formation. The prospects of the rocks of the Domanikian type as revealed through the opening of oil-wells at the Southern Tatar and Northern Tatar anticlines and other structures are quite certain but require special methods of analysis of the rocks at the wells already drilled over the terrigenous Devonian and Carboniferous deposits.

The third type of clayey reservoirs is exhibited by the rocks of the Lower Maikopian in the eastern Cis-Caucasus. They can only be tentatively related to three components because the OM in them does not occur in rock-forming proportions. According to the data from the laboratories of IG and RGI, they contain up to 3% OM. Nevertheless these rocks, consisting of clay minerals without sufficient hydrophobic fragments, contain commercial reserves of oil.

We have already mentioned the peculiar substitution of the role of OM played by silica in these rocks. It is precisely this small quantity of silica which reached the rocks during the initial stages of diagenesis that carried out the role of conservator of clay minerals, thereby lowering their exchange capacity. But this silica, committed to involvement with OM, differed from the silica precipitated from the solutions in the zone of mobilisation and could be introduced into the basin of sedimentation by the rivers. This silica of volcanogenic origin contained in its $3d$-orbitals the metals or the OM, selectively adsorbed, even though the OM occurred in very small amounts in the rocks. The groundmass of silica in the rocks of the Khadumian and Batalpashinian formations came within the second half of diagenesis, perhaps even up to the end of diagenesis, when very little water remained in the clayey part. It is well known that clayey rocks do not willingly part with water. Thus, as the silica filled up the interstices of odd configuration, it also assumed fanciful shapes.

The absence of a necessary quantity of OM in the dissolved form, sufficient to facilitate hydrophobisation of the surface of the clay particles, was reflected in the diffractograms of the rocks of the Khadumian formation (see Fig. 31). The high temperature accompanying the introduction of silica in the nearly anhydrous rocks led to its dehydration. That is why the rocks of the Khadumian and Batalpashinian formations later, while being crushed along these weakened zones at the mesolevel, gave rise to cherty surfaces reminiscent of slickensides. It is precisely this dehydration of silica at high temperatures, accompanying its discharge into the rocks of the Lower Maikopian, which exercised such an influence that the dominant clayey rocks almost without OM became reservoirs of oil. Hence, in order to forecast the prospects of a territory favourable for finding oil and allied resources, it is necessary to study in detail the history of geological development of the constituent regions through compulsory mapping of the faults and determining the time of their emplacement and appearance.

The differentiated types of clayey reservoirs require specific methods of exploitation of oil, if necessary. For rocks of the Bazhenovian type, methods of interbedded combustion should not be adopted as this might lead to elimination of the hydrophobic film over the clay minerals and concomitantly might destroy the reservoirs in this type of rocks. The technique of drilling injection wells is not recommended because infusion of water into the layers of clayey reservoirs leads to swelling of the siliceous fragments in the minerals and adversely affects the reservoir properties of the rocks. The best method is pumped-in gas although it is not completely safe because it could lead to an insignificant solution of the heavy polymers hydrophobising the surface of the clay minerals.

For the clayey reservoirs of the Domanikian type, to which the potential reservoirs of the Kuonamian and Inikanian formations belong, infusion of emulsified solutions of surfactants or pumping in gas may be recommended.

For the clayey reservoirs of the Khadumian and Batalpashinian formations the need for interformational combustion is thoroughly relevant because the increased temperature not only does not lower the reservoir properties of these rocks, but even improves them as a result of the disintegration at the mesolevel of the weakened zones formed by silica, which already possesses a certain cherty character. Under the influence of increased temperature its characteristics are improved and the rocks will yield oil more readily.

In concluding this examination of the characteristics of formation of oil reserves in clayey reservoirs, it is necessary to emphasise once again that the formation of reservoirs in clayey and clayey-carbonate rocks with a considerable content of silica and OM and the genesis of oil-pools in them is a unique process. This process belongs to katagenesis and favours vertical migration of hydrocarbons in the preliminary thinning of the rocks which prepared the ground for the accumulation of oil in them through tectonic activation of the territory where the basins of sedimentation of the clayey reservoirs were situated. The close involvement of the oil-pools with the clayey reservoirs is to be reckoned in exploitation for increased production of oil and gas in new targets.

Summary

The complex study of clayey reservoirs of the oil- and gas-bearing regions in the young and the ancient platforms as well as in the piedmont foredeeps and the additional analysis of data relating to the commercial oil reserves in the clayey reservoirs of the USA, Europe and Africa, have helped to arrive at the conclusion that the specific nature of the conditions of sedimentation and the history of tectonic development of the oil- and gas-bearing basins where clayey rocks formed were truly conducive to the holding of commercial reserves of oil- and gas so as to make them available for exploitation.

The specific characteristics of the conditions of sedimentation were facilitated by the remoteness and predominant peneplanation or sparse dissection of the relief of the region of the source rocks. Hence, a small quantity of material with a high degree of dispersion reached the basin of sedimentation. The predominance of chemical weathering in the watershed region facilitated the formation of products, primarily hydromicas in composition, from the clayey sediments along with a considerable quantity of swelling stacks. A suppression of the terrigenous source, which supplied to the basin not more than 5% of the fine silt or coarse clay of terrigenous material, is typical of the platformal areas. In the sedimentary basins of the geosynclinal type, the quantity of terrigenous minerals derived from the continent is considerably higher than in the basins of the platformal type. Further, the regions receiving a supply of sediments in the geosynclinal basins are characterised by periodic uplifts accompanied by the formation of either sandy-silty intercalations inside the thick clayey series or sandy horizons without clayey intercalations (for example, the Polyanetsian member of the Menilitovian formation).

The important characteristic of the sedimentation of the clayey sediments–the future components of the commercial reservoirs–is the presence of both allochthonous and autochthonous OM in rock-forming proportions. A distinctive role was played by OM of the third type, which hydrolysed the surface of the blocks of clay minerals and thus facilitated a decrease in their absorption capacity and increase in their filtration potential. OM of the second type also played a significant role by participating in the formation of the parallel and lenticular-layered mesotextures, because of which anisotropy of filtration indices of the rocks was produced.

One more determinative factor of sedimentation of clayey reservoirs favouring the appearance in them of high filtration potential was dispersed silicification.

While analysing the order of deposition of the silica group of minerals in the clayey reservoirs, it was found that the process of dispersed silicification was not confined to a single phase. In the three-component reservoirs, the deposition of silica took place in two stages–in the middle and late diagenesis. In the four-component clayey reservoirs, three stages of silicification in the dispersed form have been established, i.e., in the middle and late diagenesis and also in katagenesis.

Diagenesis of clay sediments was a prolonged process. It started immediately after the formation of the first portion of sediments and terminated in their lithification. At the beginning of diagenesis the sediments were considerably flooded and the more the flooding, the more the clay minerals in the rocks. Because of this, differences in the composition and concentration of salts in the silty sediments of suprabenthonic waters are practically absent. Silica reaching the basin was simultaneously and actively consumed by the planktonic organisms–the silica test builders. A gradual decrease in the quantity of water led to biochemical transformation of the OM and increased the concentration of silica in the solution. When the depositional waters entrapped within the pores attained the stage of silica saturation, the latter was precipitated. At the same time, silica was deposited not only in the form of globules of various sizes and shapes, but also and mainly in the form of a film over the mineral and organic parts of the sediments.

The property of ion exchange, determined by the degree of ionic-exchange capacity, is an important characteristic of clay minerals. In one and the same mineral, it depends upon the perfection of the crystalline structure. The degree of exchange capacity also characterised the adsorption characteristic of all other finely dispersed minerals. The clay and fine-grained carbonate minerals adsorbed the organic cations from the solutions, which led to the hydrophobisation of their surface. The adsorbed silica also hydrophobised the surface of the mineral and organic components.

The silica adsorbed by the surfaces of the microblocks of clay minerals joined them, thereby producing a hard but skeletal framework within which the clay minerals retained their plasticity and moisture, thus facilitating a decreased compactness in relation to the overlying and underlying rocks. A similar process of lesser intensity took place in the finely dispersed carbonate mass of the clayey reservoirs.

The second stage of silicification preceded the catalytic transformation of the OM adsorbed by the clay and carbonate minerals at various stages of the sedimentary process. The second stage of silicification conspicuously increased the temperature of the sediments, which stimulated the formation of new hydrocarbons from OM and the conversion of sediments into rocks. A unique redistribution of the products of transformation of OM took place. A part was adsorbed in the free pore canals or cavities formed during the solution of the part of the carbonate minerals during the transformation of OM of the second type.

The repetition of the processes of hydrophobisation of the surface of the constituent parts of the clayey reservoirs helped in the formation of complex aggregates, disposed in successive series of layers as follows: carbonate minerals–adsorbed organic products–adsorbed silica–adsorbed OM.

In a certain part of the rocks one or a few members might be deposited but hydrophobised OM or silica minerals are positively present.

The source of both dispersed and concentrated silica can be traced to endogenic impulses of its solutions into the basin of sedimentation during the onset of volcanic processes. Still, silica did not reach the sediments during their deposition, as was the case during its direct precipitation in the sediments from marine waters. Sedimentary accumulation formed layers and lenses of silicites, as seen in the sections of the Domanikian horizon and Menilitovian, Kuonamian and Inikanian formations. This is a peculiar type of accumulation of silica, i.e., in a concentrated form. Endogenic silica also stimulated the intense development of planktonic organisms, which use silica for their active life. These are mainly radiolarians, which are widely developed in clayey reservoirs.

And finally there is one more aspect of sedimentation of clayey reservoirs, the mutual disposition of the constituent parts of the sediment or, in other words, the process of formation of textural pattern which was conducive to the hydrodynamics of the basin, brought about by the character of the transporting system and also the mineral composition and the granular nature of the constituent parts of the sediments. The zones of contacts of textures of different types, described by me as weakened zones, constitute the major oil-bearing parts of all the types of clayey reservoirs. The extent and opening of these zones of contacts determine the filtration possibilities of the rocks.

The clayey reservoirs possess anisotropic filtration characteristics both parallel and perpendicular to the bedding, which are facilitated by the layers and lensoid layers of mesotextures. The possibility of migration of hydrocarbons through clayey reservoirs is guaranteed by the hydrophobisation of the contacts between the different textures of OM and silica.

The specific character of the tectonic development of the basin of sedimentation in which the future clayey reservoirs were to be formed consists in the structural differentiation, which served as the retainer of both the residual tectonic framework of earlier times and also of the imprint of the newly formed structures superimposed on the older ones. The basement faults appeared in the form of a cluster in the network of normal faults, reverse faults, flexures, overthrusts and other dislocations of such ranks, accompanied by attenuation of the closely associated rocks. In the layers of clayey rocks, the composition and textural disposition of which designate them as potential reservoirs, the components reaching them along the faults favoured their physico-chemical transformation.

The final outcome of the latter was the formation of the reservoir potential.

In the platforms the most common type of structures to which the oil formations of clayey reservoirs belong are asymmetric arches or flexures, formed in the uplifted blocks during uneven subsidences with frequent changes in the direction of movements of the different steps and blocks along the deep-seated faults. In the piedmont foredeeps, structures appearing during gravity thrust dislocations predominate.

For the formation of the commercial potential of the clayey reservoirs a tectonic activation of the structures was necessary both at the stages of sedimentation and diagenetic transformation of the sediments, and also during the completion of the stage of formation of the oil-pool. The rejuvenation of tectonic processes during the complementary stages of oil formation provided the path along which the oil-gas fluids reached the clayey reservoirs. However, the pattern ought to have been such that the frequent rejuvenation of the faults would favour only oil formation and not lead to its destruction. Hence, it is practically important not only to trace the faults, but also to determine their periods of formation and rejuvenation.

The collection of data regarding depositional and post-depositional formation of textural heterogeneity and the weakened zones in the clayey reservoirs, with specific conditions of tectonic development of the basin enabling disintegration of the weakened zones and their being filled with oil and gas due to vertical migration–all constitute major aspects of oil prospecting in sections where oil-pools exist in clayey reservoirs.

The enumerated characteristics of the palaeogeographic situation of sedimentation and tectonic development of the oil- and gas-bearing basins in the sections of which clayey reservoirs have formed, are common for all varieties of these rocks. However, the proportions of the rock-forming components and conditions of formation of the indicators of the industrial potential of the clayey reservoirs of the Domanikian horizon, Bazhenovian, Menilitovian, Khadumian, Batalpashinian, Kuonamian and Inikanian formations differ so greatly that it becomes necessary to classify them into three genetic types.

The bituminous siliceous-clayey rocks of the Bazhenovian formation of western Siberia belongs to the first type. The deposits of the Menilitovian formation of the Cis-Carpathians come under the category of a subtype. The similarity of these rocks consists in the fact that their reservoir potential arose from geochemical and structural-textural interaction of the three major components: clay minerals, OM and collomorphic silica in different degrees of recrystallisation. The textural features of these rocks at the meso- and microlevels determined the occurrence of weakened zones through which liquid and gas migrate. They were formed during the interaction of two types of OM (the first and the second) and silica with the clay minerals. The OM of these deposits consists of sapropelic and humic components. The quantity of the latter in the rocks of the

Menilitovian formation is high compared to the rocks of the Bazhenovian formation. These deposits are situated at the same stage of lithogenesis, as determined according to the predominance of the OM in them.

The conspicuous difference of the rocks of the Menilitovian formation from the deposits of the Bazhenovian formation consists in the presence mainly of parts of layers and intercalations of tuffs, tuffites, bentonite clays (in the Upper Menilitovian member) and thrice thick layers of silicites. The deposition of these rocks took place in basins of a different type. The Bazhenovian deposits accumulated in the platformal and epicontinental and the Menilitovian in the geosynclinal basins. As a result, their maximum thickness varies considerably, between 50 and 1800 m. The deposits are also differentiated according to the quantity of uranium, which increases the background values in the Bazhenovian formation by a factor of 10^2, and in the Menilitovian deposits by hardly more than 3 times, although here there is a considerable amount of humic OM, which primarily possesses the characteristic conducive for extraction of uranium from the solutions. This can be explained, perhaps, by ascribing the differences to the rates of sedimentation of the Bazhenovian and the Menilitovian formations. The duration of formation of the sediments, for example, was nearly the same–8 to 10 and 12 million years. Further, if during this period comparable amounts of uranium reached the basin of sedimentation, because of the slow rate of deposition of sediments (Bazhenov basin), the uranium tried to get completely adsorbed by the components of the rock, particularly by the humic part of the OM. During the high rate of sedimentation (Menilitov basin) uranium was distributed in amounts marginally exceeding the background value.

The conducive aspects of the clayey reservoirs of the Bazhenovian formation of the Salym oil-field, which are nowhere higher than in the studied regions of the western Siberian oil- and gas-bearing basin, were never repeated in toto. Their formation was facilitated by the specific tectonic setting of the Salym oil-field at the junction of two large blocks of the earth's crust, the textural heterogeneity leading to the appearance of weakened zones and also high temperature accompanied by tectonic activation of the territory, which helped in the hydrophobisation of the contacts as a result of thermocatalysis of OM and increased erosion of the weakened zones.

To discover new oil-pools in the clayey reservoirs of the Bazhenovian formation in the western Siberian territory, it is necessary to investigate the zones interlinked by tectonic structures of different age and character on a regional plan, over which detailed mapping of the faults of different ranks has to be carried out. This calls for a study of extensive regions favourable to new occurrences of hydrocarbons, particularly in western Siberia. From the studies conducted by the author in the Nyurol' plain region, mainly in the northern part, one can understand the objective of geoexploratory studies in the Bazhenovian formation.

The Cretaceous-Palaeogene geosynclinal oil- and gas-bearing complex, the upper stage of which comprises the Oligocene deposits (Menilitovian formation), was developed regionally along the Inner zone of the Cis-Carpathian foredeep. Nearly all the sandy-silty rocks in the complex are oil-bearing. Among the Menilitovian clayey deposits there are prospects of oil formation in the rocks of the Lower Menilitovian member. The Inner zones are broken up into a series of narrow blocks by large faults, as reflected in the overlying, particularly the Oligocene, beds in the form of a dense system of normal faults, strike-slip normal faults and, rarely, reverse or thrust faults. Such a tectonic differentiation of the territory of development of deposits of the Lower Menilitovian member and also the mineralogical and geochemical factors favourable for the formation of the commercial potential of the clayey reservoirs (presence of hydromica, enrichment of OM of the third type, hydrophobisation of the surface of contacts of microblocks of clay minerals, mesotextures with OM of the second type and silicification), highlight the prospects of the Lower Menilitovian deposits for finding oil reserves in the Inner zone of the Cis-Carpathian foredeep. One more conducive factor is the presence of oil in the sandy layer of the Oligocene situated above the deposits of the Lower Menilitovian member.

Significant exploration of the territory of the Cis-Carpathian oil and gas province should not only consist of prospecting for new oil-bearing formations, but also seek to develop existing oil-pools. In this connection, the periclinal closures intersected by the faults of the anticlinal folds and subfolded limbs of the structure complicated by faulting are particularly favourable for finding oil in the Lower Menilitovian member. To obtain commercial oil, it is necessary to probe the Lower Menilitovian deposits through the bores of open wells drilled in the territory of the Inner zone of the Cis-Carpathian foredeep wells over the Borislavian, Truskavetsian and Pokutian cover deposits.

The formation of the reservoir potential in the deposits of the Lower Menilitovian member followed the same mechanism of its formation in the rocks of the Bazhenovian formation. One of the characteristic indicators of this similarity is the nature of reflection of hydromica in diffractograms of the rocks of the Lower Menilitovian member, which exhibit a halo-forming feature with sections in the region of smaller angles and lower intensity.

The second genetic type of clayey reservoirs comprises the rocks of the Domanikian horizon of the Volya-Ural'sk region, the Kuonamian and Inikanian formations of eastern Siberia. These four-component (bituminous-clayey-siliceous-carbonate) deposits, which accumulated under similar conditions of the epicontinental basins of the platformal type, are represented by thin intercalations of rocks of varied lithological composition with a predominance of limestones, siliceous in various degrees, and also clayey with a considerable content of OM. Such a distribution of components in the rocks of the Domanikian horizon and the Kuonamian and Inikanian formations provides conditions for development of the textural inhomogeneity which defines the reservoir and filtration parameters of the rocks.

The Domanikian deposits accumulated in an isolated zone amongst the thick marine deposits of the central Frasnian basin. There were no barriers separating them. Nevertheless, they distinctly differ lithologically and faunistically from the sediments of the adjacent parts of the basin. Such a situation was complicated and hence a meridional Eastern Russian basin appearing in the Devonian was delinked from the central part of the Eastern European platform and underwent subsidence along a system of faults. The subsidence was particularly intensive during the Frasnian period, when the subsidence of the basin was not completely compensated by the sediments and is sharply reflected in the sea bottom relief. The heterogeneity of the inner structure of the basin favoured certain variations in the thickness and composition of the Domanikian deposits. Silica and microelements stimulating the rapid development of the planktonic organisms arrived along the faults in the Domanik basin. Carbonate sediments were formed in the basin in which the clayey material from the eroded source area also arrived and, together with silica and OM, created the Domanikian horizon complex.

The Kuonamian and Inikanian formations accumulated under similar conditions at the eastern part of the Cambrian epicontinental basin which, at the beginning of the Lenian period, was delinked from the western part of the narrow belt of shallow-water barrier reefs along the system of deep-seated faults. The parts of the Cambrian basin divided by the faults possessed distinctly different conditions of sedimentation. The faults delivered silica into the sediments of the Kuonamian and Inikanian formations. This silica facilitated the formation of siliceous pockets in the clayey and organic particles and also the development of organic life in the basin besides the eroded material from the continent–the clayey material with significant OM. Deposits of carbonate matter were formed in the basin and, as in the Domanikian basin, were mainly fine-grained and the OM was derived from the planktonic population. The Cambrian basin differs from the Domanikian by the presence of trilobites. In these rocks the stages of katagenesis of OM differ: in the Domanikian deposits the stage is PK-B, MK_3-Zh; in the Kuonamian formation it is PK-B_1 MK_1-D.

Finding hydrocarbons in the deposits of the Domanikian horizon, Kuonamian and Inikanian formations ought to be, first of all, directed towards mapping of faults of different orientations and determination of the deformed parts and structure of the types of flexures, normal faults, thrusts and others of analogous character present in them. For the Domanikian deposits these zones of multidirectional displacements have only been observed over the earlier faults of the basement or the deposits of the terrigenous Devonian in all the territories of the Eastern Russian basin.

The potential areas for finding hydrocarbons in the rocks of the Kuonamian and Inikanian formations are split into blocks of different forms and aligned zones of linkage of the Vilyuian syneclise with the terminal parts of the Pri-Verkhoyan depression. The depths in which these deposits are situated vary from 5–6 km and are accessible to modern drilling techniques and might be taken up for detailed prospecting. Further, in the limbs of these structures the depths would be considerably less, thus making oil geological management more feasible.

The rocks of the Khodzhaipakian formation of the Upper Jurassic of Central Asia resemble those of the Kuonamian formation in composition and conditions of formation. The tectonic setting of the basin in which the rocks of the Khodzhaipakian formation were formed was unique and the structure of its floor inherited that of both the basement and the overlying rocks of the Kellovian-Oxfordian. The relief of the floor determined the location of the reefs and sediments, which filled the depressions. Parts of the basin where these types of sediments accumulated were bound by faults that supplied the ingredients for both the reef-builders and the clayey-carbonate sediments, inducing high gamma activity. Sedimentation of fine-grained carbonate material and planktonic OM took place in the basin. Clay minerals and OM of the first and second types arrived from the land. The rocks of the Khodzhaipakian formation are not exactly similar to those of the Domanikian type because the silica in them never attained rock-forming proportions.

Prospective regions for finding oil and gas fluids in the rocks of the Khodzhaipakian formation include all the fields in which, at the present time, exploitation of commercial oil and gas is carried out from the reefs and which, like the oil-fields of Kultak, are situated close to the regional faults bound by structures of varying nature. These oil formations ought to be analysed for determination of the productivity of the rocks of the Khodzhaipakian formation. It is only necessary to adhere strictly to the inherent conditions while testing the clayey reservoirs. Although the rocks of the Khodzhaipakian formation do not possess high industrial parameters, they ought to be considered standby reserves, particularly when secondary exploitation of the formations is undertaken in future.

The rocks of the lower part of the Maikopian series (Khadumian and Batalpashinian formations) belong to the third genetic type of clayey reservoirs. In fact, these are two-component (siliceous-clayey) rocks in which the OM does not occur in rock-forming proportions. Because of the characteristics of its structures, silica played the role of a peculiar substitute for OM in the rocks of the Khadumian and Batalpashinian formations. The $3d$-orbitals were either free or could adsorb the OM present in the rocks only in minimum quantities or could take up OM or metals with the formation of organo-siliceous or organo-metallic bonds possessing high strength, to be uniformly adsorbed by the surface of the clay minerals and form in them a protective hydrophobic film. In spite of its smaller amounts, the OM contains a high percentage of chloroform bitumenoids. According to thermoanalytical data, the OM of the rocks of the Khadumian formation consists of euxines, spores and cuticles with predominant euxinite. It is found in the lower stage of transformation. In the rocks of the Batalpashinian formation, the OM consists mainly of vitrinite. The OM of these two formations was adsorbed by clay minerals and silica and also played a minor role in the formation of mesotextures.

The addition of silica in the rocks of the Maikopian series took place at the close of diagenesis. Silica occupied places where, at that time, water still remained. The various modes of distribution of the silica adsorbed by the clay minerals was responsible for the diversity of forms of dislocations exhibited by these rocks and the weakened parts of the rocks at the mesolevel. The weakened zones served as the major storage space in the rocks and also the main pathways for the migration of oil. The absence of radiolarians is another indicator of the fact that the entry of silica into the rocks of the Khadumian and Batalpashinian formations belonged to the second half of diagenesis and was not synsedimentary. The time of silicification of the rocks of the Khadumian and Batalpashinian formations greatly influenced the formation of their reservoir potential in such a way that these rocks differ from those of the Domanikian horizon, Bazhenovian, Kuonamian and Inikanian formations and also the Lower Menilitovian member. Silica entered these rocks at the beginning of diagenesis and formed a rigid skeletal framework within which the clayey matter retained plasticity and moisture. In the rocks of the Khadumian and Batalpashinian formations, no silica network formed but zones of dislocations of various forms of considerable length (1.2–3.5 mm) and width (0.04–0.08 mm) appeared along which the rocks split, as evidenced by the presence of slickensides.

The prospects of striking oil in the clayey reservoirs of the Khadumian and Batalpashinian formations can be related only to the contacts of structures of different nature, on whose borders under the influence of basement faults various types of dislocation structures would form (flexures, tight limbs of anticlines, gravity faults and others). The actual problem in the eastern Cis-Caucasus during oil exploration in the clayey reservoirs is the mapping of the basement faults and a search for structures in the sedimentary cover initiated by them.

The clayey reservoirs of the Pilengian formation from the Miocene of Sakhalin Island belong to this genetic type. Their similarity to the rocks of the Khadumian formation is as follows: The same nature of changes in ratio of clay minerals to silica is observed throughout the section; their OM does not occur in rock-forming quantities; similarly, the quantity of terrigenous admixture varies along the section; and oil reserves are present. The difference lies in the time of arrival of silica in the sediments. Both in the rocks of the Khadumian and Batalpashinian formations as well as in the rocks of the Pilengian, there were two stages of addition of silica to the sediments, but the timings of its generation and incorporation were different. In the rocks of the Khadumian formation, silica reached during the middle and at the close of diagenesis and during both stages was injected only as a chemical precipitate. Its quantity was also not uniform as the major amount reached the sediments at the final stage of diagenesis. In the rocks of the Pilengian formation, the genetic types of silica were equal contributors. While one type was precipitated from solutions, the other was biogenic. The former's entry into the sediments took place at the beginning of diagenesis, while the latter was reorganised within the sediments only chemically in the middle of diagenesis.

The reservoir capacity of the third genetic type was formed at the expense of the weakened zones at meso- and microlevels, but in the rocks of the Pilengian formation, there appeared an additional capacity due to the fractures which developed during the dehydration of silica. In the rocks of the Khadumian formation, the dislocations are of a different type.

To find oil and gas in this region the rocks which are equivalent to those in the Pilengian formation are prospective but possess different names in the different sedimentary basins of the northwestern part of the Pacific Ocean mobile belt. The necessary conditions for the prospective reserve potential of a deposit of such a type are the tectonic activation of the territory at the moment of formation of the deposits and the fault tectonics at the time of formation of the reservoir and at the time of migration of hydrocarbons.

The potential reservoirs of the lower Permian of the Pri-Caspian basin cannot be relegated in toto to any one of the types described. Their principal rock-forming components are clay minerals, organic matter and microgranular carbonate-calcite, which sometimes assumed rock-forming proportions. There is no admixture of terrigenous minerals. Mineralogical and textural-structural transformation of the rocks mainly favoured the compaction related to the great depths of occurrence of the rocks of clayey composition. The potential clayey reservoirs of the Lower Permian have either layered or massive mesotextures. The microtextures are lumpy and enclosing. In providing the useful reservoir capacity for the clayey reservoirs of the Lower Permian, in addition to weakened zones, tectonic fractures, which formed due to active disjunctive tectonics, also contributed during the restructuring of the territory of their distribution.

The wide development of disjunctive tectonics and also the commercial oil formation in the underlying and overlying deposits make us think that the potential clayey reservoirs of the Sakmaro-Artinian part of the Lower Permian are a promising productive zone of oil reserves. Another significant feature in favour of these rocks as prospective hydrocarbon sources is the fact that structures of different origin formed far earlier than the commencement of migration of hydrocarbons from their places of formation in the reservoirs of the subsaline and suprasaline deposits. The unique genetic type of oils in both the complexes has been revealed in geochemical studies. The formation of oil-pools through vertical migration has also been emphasised by the results of palaeophytological studies.

The composition and textural features of the rocks of the Lower Permian Pri-Caspian basin, characterised by the absence of silica and the presence of minor amounts of OM, compel us to suggest that these would be reservoir rocks of modest filtration parameters.

The potential clayey reservoirs are most prospective in those parts of the territory of their occurrence where the regional fault separates structures of differing character, as in the Kenkiyak zone, but the prospective zones would nevertheless be situated within the plumage faults.

The textural heterogeneity at the meso- and microlevels, which is responsible for the appearance of the weakened zones at the contacts of textures of various types, serves as the main cause of formation of the reservoir capacity of clayey reservoirs. However, the filtration possibilities of clayey reservoirs were realised only at the moment of their being filled with oil through vertical migration. Vertical migration is the principal mechanism whereby clayey reservoirs were filled with oil and gas because the special feature of the reservoir space in such reservoirs–the weakened zones–required considerable energy for their opening. Hence, lateral migration along the bedding of the clayey reservoirs was possible only at a very limited distance from the source of migration, the faults. One proof of the formation of clayey reservoirs during vertical migration of hydrocarbons is the fact that in all the oil occurrences where oil has been found in such reservoirs, it has also been present in both the overlying and underlying deposits.

Extreme care must be exercised in the choice of methods of intensification of oil production in relation to the specific features of formation of a reservoir, namely, the filtration properties of clayey reservoirs. The method of intensification should be selected in accordance with the characteristics of a given reservoir.

Intralayered combustion should be avoided in the case of rocks of the Bazhenovian formation because this would lead to the destruction of the hydrophobic film or, in other words, the reservoir. Forceful injection of water into the layers should also be avoided because this would lead to the swelling of siliceous fragments in the clayey and organic particles and, consequently, to the degradation of the reservoir properties of the rocks. Pumping in gas could be adopted but, though administered in small degrees, it might lead to the destruction of the hydrophobic film. For the rocks of the Domanikian horizon, the use of emulsified solutions of surfactants or pumping in gas is recommended. As for the clayey reservoirs of the Lower Maikopian, resorting to intralayered combustion is highly relevant here because increased temperature not only does not impair, but even considerably enhances their reservoir potential by opening up the weakened zones at the mesolevel.

Literature Cited*

1. Gurari, F.G., V.M. Gavshin, N.I. Matvienke *et al.* 1984. Assetsiatsiya mikroznementev s organicheskim veshchestvom v osadochnykh tonshchakh Sibiri [Association of Microelements with Organic Matter in the Sedimentary Deposits of Siberia]. Izd. SNIIGIMSa, Novosibirsk.
2. Bakhturov, S.F. and B.S. Pereladov. 1984. Stroenie i usloviya obrazovaniya kuonamskoi svity vostoka Sibirskoi platformy [Structure and conditions of formation of the Kuonamian formation east of the Siberian platform]. In: *Osadochyne Formatsii i Usloviya Ikh Obrazovaniya.* Novosibirsk, pp. 79–99.
3. Biktyasheva, R.M. 1980. Regional 'naya neftenosnost' verkhnedevonsko-turneiskogo karbonatnogo kompleksa Tatarii [Regional oil-bearing nature of the Upper Devonian-Turneiian carbonate complex of Tatar]. *Geologiya Nefti i Raza*, no. 4, pp. 51–55.
4. Bondarik, G.K., A.M. Tsareva and V.V. Ponamarev. 1975. Tekstura i deformatsiya glinistykh porod [Texture and Deformation of Clayey Rocks]. Nedra, Moscow.
5. Van, A.V. 1984. Model' vunkanogenno-osadochnogo obrazovaniya domanikitov [Model of volcanogenic sedimentary formation of the Domanikites]. In: *Problemy Geologii i Neftegazonosnosti Verkhnepaleozoiskikh Otlozhenii Sibiri.* Novosibirsk, pp. 79–87.
6. Veber, V.V. and L.A. Kotseruba. 1979. Usloviya bituminoznosti bazhenovskoi svity Zapadnoi Sibiri [Conditions of bituminous deposition in the Bazhenovian formation of western Siberia]. *Geologiya Nefti i Gaza,* no. 9, pp. 15–19.
7. Gabinet, M.P. 1985. Postsedimentatsionnye preobrazovaniya flisha Ukrainskikh Karpat [Post-sedimentation Transformations of the Flysch of the Ukrainian Carpathians]. Naukova Dumka, Kiev.
8. Gaideburova, E.A. 1982. Tipy razrezov domanikitov Zapadnoi Sibiri [Types of sections of Domanikites in western Siberia]. In: *Domanikity Sibiri i Ikh Rol' v Neftegazonosnosti.* Novosibirsk, pp. 23–32.
9. Kontorovich, A.E., I.I. Nesterov, F.K. Salmanov *et al.* 1975. Geologiya nefti i raza Zapadnoi Sibiri [Geology of Oil and Gas in Western Siberia]. Nedra, Moscow.

* Some references were incomplete in the Russian original—Language Editor.

10. Gritsaenko, G.S. and M.I. Il'in. 1975. Rastrovaya elektronnaya mikroskopiya mineralov [Scanning electron microscopy of minerals]. *Izv. AN SSSR, Ser. Geol.*, no. 7, pp. 24–34.
11. Gurari, F.G. 1984. Regional'nyi prognoz promyshlennykh skoplenii urlevodorodov v domanikitakh [Regional prognosis of the commercial accumulations of hydrocarbons in the Domanikites]. *Geologiya Nefti i Gaza,* no. 2, pp. 1–5.
12. Aliev, M.M., G.P. Batanova, R.O. Khachatryan *et al.* 1978. Devonskie otlozheniya Volgo-Ural'skoi neftegazonosnoi provintsii [Devonian Deposits of the Volga-Ural'sk Oil and Gas-bearing Province]. Nedra, Moscow.
13. Dmitrievskii, A.N. 1982. Sistemnyi litologogeneticheskii analiz neftera-zonosnykh osadochnykh basseinov [Systems Analysis of the Lithology and Genesis of Oil and Gas-bearing Sedimentary Basins]. Nedra, Moscow.
14. Dobrynin, V.M. 1982. Problema kollektora nefti v bituminoznykh glinistykh porodakh bazhenovskoi svity [Problem of oil reservoir in the bituminous clayey rocks of the Bazhenovian formation]. *Izv. AN SSSR, Ser. Geol.*, no. 3, pp. 120–127.
15. Gurari, F.G. [ed.] 1982. Domanikity Sibiri i ikh rol' v neftegazonosnosti [Domanikites of Siberia and Their Role in Oil and Gas Formation]. Trudy SNIIGGIMSa, Novosibirsk.
16. Zaripov, O.G. and I.I. Nesterov. 1977. Zakonomernosti razmeshcheniya kollektorov v glinistykh otlozheniyakh bazhenovskoi svity i ee vozrastnykh analogov v Zapadnoi Sibiri [Pattern of disposition of reservoirs in the clayey deposits of the Bazhenovian formation and their temporal analogues in western Siberia]. *Sovetskaya Geologiya*, no. 3, pp. 19–26.
17. Pararova, G.M., C.G. Nerichev, A.I. Ginzburg *et al.* 1984. Iskhodnyi material i fatsial'no-geokhimicheskie usloviya formirovaniya veshchestvenno-petrograficheskogo sostava organicheskogo veshchestva raznovozrastnykh domanikoidnykh otlozhenii [Original material and facies—geochemical conditions of formation of the petrographic composition of the multi-age organic matter of the Domanikian deposits]. *Geokhimiya*, no. 12, pp. 1882–1885.
18. Savitskii, V.E., V.M. Evtushenko, L.I. Egorova *et al.* 1972. Kembrii Sibirskoi platformy [The Cambrians of the Siberian Platform]. *Trudy SNIIGGIMSa,* Novosibirsk, no. 130.
19. Klubova, T.T. 1973. Glinistye mineraly i ikh rol' v genezise, migratsii i ak kumulyatsii nefti [Clayey Minerals and Their Role in the Genesis, Migration and Accumulation of Oil]. Nedra, Moscow.
20.* Klubova, T.T. 1980. Osobennosti migratsii nefti cherez glinisto-karbonatnye porody [Characteristics of migration of oil through clayey

* Not referenced in text—Language Editor.

carbonate rocks]. In: *Porody-kollektory i Migratsiya Nefti*. Nauka, Moscow, pp. 92–97.

21. Dorofeeva, T.V. [ed.] 1983. Dollektory nefti bazhenovskoi svity Zapadnoi Sibiri [Reservoirs of Oil in the Bazhenovian Formation of western Siberia]. Nedra, Leningrad.
22. Konysheva, R.A. and A.P. Roznikova. 1977. Stroenie otlozhenii bazhenovskoi svity Salymskoi ploshchadi po dannym rastrovoi zlektronnoi mikroskopii i vychislitel'nogo ustroistva Kvantimet-720 [Structure of the deposits of the Bazhenovian formation of the Salym oil-field according to data from scanning electron microscopy and the computer facility Quantimet-720]. In: *Problemy Geologii Nefti*. Moscow, pp. 76–89.
23. Konysheva, R.A. and R.S. Sakhibgareev. 1976. O prirode emkosti v argillitakh bazhenovskoi svity Zapadnoi Sibiri [On the nature of reservoir capacity in the argillites of the Bazhenovian formation of western Siberia]. *Dokl. AN SSSR*, 228, 5, 1197.
24. Kosarev, V.S. 1985. O neftenosnosti nizhnemaikopskikh otlozhenii Vostochnogo Stavropol'ya [On the oil-bearing nature of the lower Maikopian deposits of eastern Stavropol']. In: *Nauchnye Osnovy Poiskov i Razvedki Neftyanykh Mestorozhdenii*. Moscow, pp. 119–128.
25. Krylov, N.A., B.V. Kornev and M.I. Kozlova. 1978. Osobennosti razmeshcheniya zalezhei nefti bazhenovskoi svity v raionakh Srednego Priob'ya [Characteristics of deposition of oil reserves in the Bazhenovian formation in the central Pri-Ob' region]. *Trudy IG i RGI*, Moscow, no. 16, pp. 44–54.
26. Matvienko, N.I. and V.I. Moskvin. 1982. Osobennosti geokhimicheskoi sredy nakopleniya goryuchikh slantsev Yakutii [Characteristics of the geochemical medium of accumulation of combustible shales of Yakutiya]. In: *Litologiya Rezervuarov Nefti i Gaza v Mezozoiskikh i Paleozoisk Ikh Otlozheniyakh Sibiri*. Novosibirsk, pp. 40–45.
27. Messezhnikov, M.S. 1983. Paleogeografiya i biostratigrafiya yury i mela Sibiri [Palaeogeography and biostratigraphy of the Jurassic and Cretaceous in Siberia]. *Trudy IG i G CO AN SSSR*, Novosibirsk, no. 528, pp. 32–46.
28. Nesterov, I.I. 1980. Neftegrazonosnost' bituminozhykh glin bazhenovskoi svity Zapadnoi Sibiri [Oil-bearing nature of the bituminous clays of the Bazhenovian formation of western Siberia]. *Sovetskaya Geologiya*, no. 11, pp. 3–10.
29. Dolenko, G.N., L.T. Boichevskaya, M.B. Boichuk *et al.* 1985. Neftegazonosnye provintsii Ukrainy [Oil and Gas-bearing Provinces of the Ukraine]. Naukova Dumka, Kiev.
30. Krylova, N.A. [ed.] 1980. Neftenosnost' bazhenovskoi svity Zapadnoi Sibiri [Nature of the Oil-bearing Bazhenovian Formation of Western Siberia]. Izd. IG i RGI, Moscow.

31. Onishchenko, B.A. 1986. Uspoviya osadkonakonleniya i neftegazonosnost' maikonskikh otlozhenii Predkavkaz'ya [Conditions of sedimentation and nature of oil and gas-bearing Maikopian deposits of the Cis-Caucasus]. *Geologiya Nefti i Gaza*, no. 2, pp. 23–27.
32. Chepak, G.N., B.M. Shaposhnikov, P.S. Naryzhnyi *et al.* 1983. Osobennosti neftenosnosti glinistoi tolshchi oligotsena Predkavkaz'ya [Characteristics of the oil-bearing nature of the Oligocene clayey depositions in the Cis-Caucasus]. *Geologiya Nefti i Gaza*, no. 8, pp. 36–40.
33. Mirchink, M.F., O.M. Mkrtchyan, A.A. Trokhova *et al.* 1975. Paleotektonicheskie i paleogeomorfologicheskie osobennosti Volgo-Ural'skogo domanikovoro basseina [Palaeotectonic and Palaeogeomorphological Characteristics of the Volga-Ural'sk Domanik Basin] *Izv. AN SSSR, Ser. Geol.*, no. 12, pp. 9–18.
34. Pluman, I.I. 1975. Rasprostranenie urana, toriya i kaliya v otlozheniyakh Zapadno-Sibirskoi plity [Distribution of uranium, thorium and potassium in deposits of the Western Siberian plate]. *Geokhimiya*, no. 5, pp. 756–766.
35. Pluman, I.I. and N.P. Zapivalov. 1977. Usloviya obrazovaniya bituminoznykh argillitov volzhskogo yarusa Zapadno-Sibirskoi neftegazonosnoi provintsii [Conditions of formation of bituminous argillites of the Volzh belt of western Siberian oil and gas-bearing province]. *Izv. AN SSSR, Ser. Geol.*, no. 9, pp. 111–117.
36. Reifman, L.M., Yu. Kh. Ovcharenko and M.A. Vul'. 1985. Vydelenie markiruyushchikh gorizontov v oligotsene Karpat pri pomoshchi gamma-karotazha [Delineation of marker horizons in the Oligocene of the Carpathians using gamma-logging]. *Dokl. AN UkrSSR, Ser. B. Geologiya, Khimiya i Biol. Nauki*, no. 5, pp. 30–33.
37. Sokolova, M.F. 1976. Nekotorye itogi sinteza gidroslyudistykh mineralov [Some results on the synthesis of hydromicaceous minerals]. In: *Modelirovanie i Fiziko-khimiya Litogeneza*. Novosibirsk, pp. 82–91.
38. Vysotskogo, I.V. [ed.] 1976. Spravochnik po neftyanym i gazovym mestorozhdeniyam zarubezhnykh stran. Kniga l. Evropa, Severnaya i Tsentral'naya Amerika [Handbook on Oil and Gas Depositions in Foreign Countries. Book 1. Europe, North and Central America]. Nedra, Moscow.
39. Stepanov, A.I. and Yu. A. Tereshchenko. 1985. Tip kollektora i uspoviya formirovaniya zalezhi nefti v otlozheniyakh bazhenovskoi svity Salymskogo mestorozhdeniya [Type of reservoir and conditions of formation of oil reserves in deposits of the Bazhenovian formation in the Salym locality]. In: *Neftegazopromyslovaya Geologiya Zalezhei s Trudnoizvlekaemymi Zapasami*. Moscow, pp. 80–93.
40. Stroenie i i neftegazonosnost' bazhenitov Zapadnoi Sibiri [Structure and nature of the bazhenites of western Siberia]. *Trudy ZapSibNIGNI, Tyumen'*, no. 175.

41. Taruts, G.M. and E. A. Gaideburova. 1978. Stroenie neftegazonosnykh otlozhenii bazhenovskoi svity Zapadno-Sibirskoi plity v svyazi s osobennostyami tektoniki verkhneyurskogo basseina osadkonakopleniya na primere Salymskogo raiona [Structure and nature of oil and gas-bearing deposits of the Bazhenovian formation in the Western Siberian plate in relation to the tectonic characteristics of sedimentation in the Upper Jurassic basin, for example in the Salym region]. In: *Tsiklichnost' Neftegazonosnykh Basseinov i Zakonomernosti Razmeshcheniya Zalezhei.* Novosibirsk, pp. 69–80.

42. Filina, S.I., M.V. Korzh and M.S. Zonn. 1984. Paleogeografiya i neftenosnost' bazhenovskoi svity Zapadnoi Sibiri [Paleogeography and Nature of the Oil-bearing Bazhenovian Formation of Western Siberia]. Nauka, Moscow.

43. Khabarov, V.V., O.V. Bartashevich and O.M. Nelepchenko. 1981. Geologo-geofizicheskaya kharakteristika i neftenosnost' bituminoznykh porod bazhenovskoi svity Zapadnoi Sibiri [Geological and geophysical characteristics and nature of the oil-bearing bituminous rocks of the Bazhenovian formation of western Siberia]. *Obzor. Ser. Geologiya, Metody Poiskov i Razvedki Mestorozhdenii Nefti i Gaza.* VIEMS, Moscow.

44.* Khalimov, E.M. and V.S. Melik-Pashaev. 1980. O poiskakh promyshlennykh skoplenii nefti v bazhenovskoi svite [On explorations of industrial accumulations of oil in the Bazhenovian formation]. *Geologiya Nefti i Gaza*, no. 6, pp. 1–10.

45. Klubova, T.T. 1985. Pattern of interactions of clay minerals and organic matter. In: *Reports of 5th Meeting of European Clay Groups* [ed. J. Konta]. Prague, pp. 253–257.

46. McCaslin, J.C. 1981. Colorado's Niobrara still holds interest. *Oil and Gas J.*, 79, 49, 231–233.

47. McCaslin, J.C. 1980. San Luis and San Juan basins draw interest. *Oil and Gas J.*, 78, 40, 111.

48. Rodrigues, R., L.P. Quadros and A.L. Scofild. 1979. Caracterizacao da materia organica das rochas sedimentares poz analise termica diferencial e termogravimetrica. *Boletin Tecnico da Petrobras*, 22, 1, 3–20.

49. Truex, J.N. 1972. Fractured shale and basement reservoir, Long Beach Unit, California. *Bull. Amer. Assoc. Petrol. Geol.*, 56, 10, 1931–1938.

50. Van Tyne, Arthur M. 1981. World energy developments. *Bull. Amer. Assoc. Petrol. Geol.*, 65, 10, 1801–1813.

* Not referenced in text—Language Editor.

Printed in India